Introduction to
CRYSTAL GROWTH
Principles and Practice

Introduction to
CRYSTAL GROWTH
Principles and Practice

H. L. Bhat

Former Professor of Physics
Indian Institute of Science
Bangalore

and

Visiting Professor
Centre for Nano and Soft Matter Sciences
Bangalore

CRC Press
Taylor & Francis Group
Boca Raton London New York

CRC Press is an imprint of the
Taylor & Francis Group, an **informa** business

CRC Press
Taylor & Francis Group
6000 Broken Sound Parkway NW, Suite 300
Boca Raton, FL 33487-2742

First issued in paperback 2016

Version Date: 20150126

ISBN 13: 978-1-138-19971-2 (pbk)
ISBN 13: 978-1-4398-8330-3 (hbk)

Library of Congress Cataloging-in-Publication Data

Bhat, H. L., author.
  Introduction to crystal growth : principles and practice / H.L. Bhat.
    pages cm
  Includes bibliographical references and index.
  ISBN 978-1-4398-8330-3 (hardcover : alk. paper) 1. Crystal growth. I. Title.

QD921.B53 2015
548'.5--dc23   2014039019

Visit the Taylor & Francis Web site at
http://www.taylorandfrancis.com

and the CRC Press Web site at
http://www.crcpress.com

To my parents

Contents

Section II Practice

Preface

"Yet another book on crystal growth? Is it really necessary?" These questions were always in the back of my mind. The motivation, however, was that I had prepared extensive notes on this topic for many years while teaching the subject to graduate students who would eventually do some crystal growth before investigating their physical properties. Furthermore, there was constant encouragement and persuasion from the publishers, particularly Ms. Aastha Sharma, without which I doubt this book would have been written.

The treatment of this book is at an introductory level, as the title suggests. It is intended for those curious students who have developed some fascination for crystals and hence want to learn more about them and their origin. Accordingly, some historical perspective of the subject is initially given and wherever possible some background information is included. In my view, knowing the growth of the subject is as important as knowing the recent developments. Most of the ideas presented here have been borrowed from many authoritative books and review articles, and wherever possible due credits have been given to their authors.

Many persons have helped me in the preparation of this book. First of all, I would like to thank my colleague, Professor G. S. Ranganath, who read the entire manuscript critically as an "outsider." I have greatly benefited from the many discussions I had with him. I would like to express my gratitude to my other colleague, Professor K. A. Suresh, for his constant encouragement and support. I also want to thank my colleagues at the Indian Institute of Science, Bangalore, Dr. Suja Elizabeth and Professor P. S. Anilkumar for their kind support.

I am very grateful to my research students and others who have worked with me during the past four decades. Their contributions have strengthened the content of this book. In particular, I would like to thank K. Ragahvendra Rao, who has helped me a great deal in preparing some artwork and photographs as well as work related to copyright permissions and references. I would also like to thank all the publishers for their kind permission to reproduce certain figures in this book.

Finally, I acknowledge the debt I owe to my wife, Prema, who has put up with me during the difficult days of writing this book. Her constant encouragement and support were vital for the successful completion of this work.

H. L. Bhat

Author

Prof. H. L. Bhat obtained his BSc degree from Mysore University and both MSc and PhD degrees from Sardar Patel University, Gujarat, India. He obtained his postdoctoral research training from Strathclyde University, Glasgow, United Kingdom. Dr. Bhat joined the Physics Department of Indian Institute of Science, Bangalore, as a research associate in 1973, and he progressed steadily to become a professor in 1993. He was the chairman of this department during 2002–2006. During 2006–2008, he was a CSIR Emeritus Scientist. Currently, he is a visiting professor at the Centre for Nano and Soft Matter Sciences, Bangalore. His research interests include crystal growth and crystal physics with special reference to ferroelectric, nonlinear optical, semiconductor, and magnetic materials. He has published 219 research articles and produced 28 PhD candidates. He is a member of many professional bodies and is currently the joint secretary for the Materials Research Society of India.

Section I

Principles

1

Introduction

What Are Crystals?

Matter exists in three states—gaseous, liquid, and solid—ignoring the fourth state called the plasma state, which consists of equal number of positive ions and electrons. These three general states of matter represent very different degrees of atomic or molecular mobility. In the gaseous state, the molecules are in constant, vigorous, and random motion. A gas let into an evacuated container takes the shape of the container, is readily compressible, and exhibits very low viscosity. In the liquid state, the random molecular motion is much more restricted. Unlike a gas, the volume occupied by a liquid is limited. The liquid only takes the shape of the occupied part of its container, and its free surface is flat except in regions where it comes into contact with the container walls. A liquid exhibits a much higher viscosity than a gas and is less easily compressed. In the solid state, however, molecular motion is confined to an oscillation about a fixed position, and the structure is elastically deformable only to a small extent. In fact, it will often fracture when subjected to a large deforming force.

Solids may be crystalline or amorphous, and the crystalline state differs from the amorphous state in the regular arrangement of constituent molecules, atoms, or ions into some fixed and rigid pattern known as a lattice. The term "crystalline" is very frequently used to indicate the high degree of internal regularity resulting in the development of definite external faces. Hence, crystals grown without space constraints are polyhedral, bounded by flat faces meeting at straight edges. These flat faces, which are highly polished in good crystals, are inclined to each other at angles that are characteristic of a particular material irrespective of the external shape. The interfacial angles display certain symmetries and are in accordance with the structural symmetries possessed by the crystal. In fact, accurate measurement of the interfacial angles has shown that often there exists a small variation in interfacial angles from one crystal to another of the same substance, and these variations were discovered even between the faces giving sharp light signals in the optical goniometer. It was found that the observed interfacial angles differed from the theoretical values calculated on the basis of law of rational indices by 30–60 seconds. This shows that the main natural faces of crystals make a very small and variable angle with low-index planes. Because these faces are very near to the faces of simple forms, they possess high Miller indices. Such faces are called vicinal faces. We shall refer to the importance of these observations in Chapter 5.

Crystallographers call the characteristic shape of a crystal its habit. However, within the same habit, the relative sizes of the faces of a particular crystal may vary considerably. A given crystal may grow rapidly or have stunted growth in one of the crystallographic directions. Thus, an elongated growth of a prismatic crystal gives a needle-shaped crystal (acicular habit). Likewise, stunted growth in one particular direction gives a flat, plate-shaped crystal (tabular habit). This kind of variation is called habit modification.

Why Single Crystals?

Single crystals find important uses in research and development. In fact, according to one classification, all true solids are crystals. Thus, for an understanding of the physics and chemistry of the solid state, single crystals are a prerequisite. One may use polycrystalline samples rather than single crystals for many studies, but often single crystals are preferred. As we know, polycrystals contain grain boundaries. If we desire to know some bulk property of a material and we measure that property on a polycrystalline specimen, it will in many cases include the effect of grain boundaries. A notable example of a property where single crystals are essential is electrical conductivity, which is particularly impurity sensitive. Impurities tend to segregate at grain boundaries, thereby obstructing the passage of current. Hence, single crystals are almost always required for determining any conductivity-dependent property. Another common effect of grain boundaries and associated voids is light scattering and, hence, single crystals are required in optical studies as well.

Many properties of crystals depend on the crystallographic direction in which the measurement is carried out, because the spatial arrangement of the constituent atoms is not in general the same in all directions. Consequently, if we determine a directionally dependent property in a polycrystalline specimen, where the crystallites are randomly oriented, we will obtain only an average value of the property in which the directional dependence gets masked.

Single crystals have important practical applications in technology. For example, in quartz oscillators, much better frequency stability can be achieved in single crystal quartz than polycrystalline samples. In nonlinear optical conversion processes, one invariably uses oriented single crystals. One can give many such examples where single crystals are always required. Some of the important applications that single crystals find in technology and device research are given in Table 1.1.

TABLE 1.1

Practical Applications of Single Crystals

Sl. No.	Material Class	Devices	Crystals Used
1	Semiconductors	a. Electrical diodes	Si, Ge
		b. Hall effect magnetometer	InSb
		c. Integrated circuits	Si, GaAs
		d. Infrared detectors	GaSb, InAs, $Cd_xHg_{1-x}Te$
		e. Light-emitting diodes	GaAs, GaSb, GaP, $Sn_xPb_{1-x}Te$, GaN, ZnTe
		f. Photo diodes	Si, GaAs, $Cd_xHg_{1-x}Te$
		g. Photo conduction devices	Si, $Cd_xHg_{1-x}Te$
		h. Radiation detectors	Si, Ge, CdTe, BGO, PbS
		i. Transistors	Ge, Si, GaAs
		j. Thyristors	Si
2	Optical materials	a. Electro-optic devices	$LiNbO_3$, ADP, KDP
		b. Laser hosts	YAG, $Al_2O_3:Cr^{3+}$ (ruby), alexandrite, $CaWO_4$, $Ti:Al_2O_3$, GaAs, AlGaAs, InP, InSb, AlGaInP, InGaAsSb, InGaN
		c. Lenses, prisms, windows	Al_2O_3 (0.15–55 µm) Ge (1.8–22) LaF_3 (0.4–11 µm) Si (1.2–15 µm) CaF_2 (0.12–10 µm) MgF_2 (0.11–8 µm) AgBr (0.5–35 µm) AgCl (0.4–28 µm) LiF (0.12–7 µm) CsBr (0.23–45 µm) Quartz (0.19–4 µm) ZnS
		d. Magneto-optic devices	YIG ($Y_3Fe_5O_{12}$)
		e. Nonlinear optical devices	ADP, KDP, $LiNbO_3$, KTP, BBO, LBO, CLBO
		f. Polarizers	$CaCO_3$, $NaNO_3$
		g. x-Ray monochromators	Si, KAP
		h. Photorefractive devices (holographic data storage, phase conjugate mirrors)	$BaTiO_3$, $LiNbO_3:Fe^{3+}$, BSO ($Bi_{12}SiO_{20}$)
		i. Scintillation detectors	NaI:Tl, BGO ($Bi_4Ge_3O_{12}$)
3	Hard materials and materials for mechanical components	a. Abrasives and cutting tools	SiC, diamond, sapphire (Al_2O_3)
		b. Bearings	Al_2O_3, Si_3N_4, ruby ($Al_2O_3:Cr^{3+}$)
		c. Substrates for high T_c superconductors	$SrTiO_3$
		d. Strain gauges	Si, Ga(As, P)
		e. Cantilevers	Si, Si_3N_4

Continued

TABLE 1.1 (*Continued*)

Practical Applications of Single Crystals

Sl. No.	Material Class	Devices	Crystals Used
4	Piezoelectric materials	a. Resonant bulk wave devices	SiO_2, $LiTaO_3$
		b. Surface wave devices	SiO_2, $LiNbO_3$
		c. Transducers	Quartz, Rochelle salt, ADP, etc.
5	Magnetic materials	a. Microwave filters	Garnets ($Y_3Fe_5O_{12}$, $Y_3Al_5O_{12}$)
		b. Tape heads	Ferrites
6	Pyroelectric materials	Pyroelectric devices	TGS, $LiTaO_3$ $Ba_xSr_{1-x}Nb_2O_6$ (fire alarm)
7	Gems	Jewelry	Cubic zirconia (ZrO_2), diamond, ruby, sapphire, etc.

Why Should One Grow Crystals?

From Table 1.1, it is clear that single crystals are used in a wide variety of applications, and the global requirement for these crystals is continuously growing. For these applications, the crystals to be used should meet certain minimum specifications in terms of purity and perfection. Not all the required crystals are available in nature. Even those that are available in nature may not be of the purity level required for certain devices. Furthermore, many of the technologically and commercially important crystals are all synthetic. Hence, growing crystals under controlled conditions has become a necessity. Although development and synthesis of newer crystals often happen in academic institutions, large-scale production of important and useful crystals is mainly being undertaken by industries. In fact, crystal growth has now become a multi-billion-dollar industry.

Historical Perspective

Crystal growth is a very old subject. The historical development of this subject is fascinating, and it involved many people across the world with diverse expertise. In this section, we dwell on it only briefly and base our discussion on the chapter written by Scheel in the *Handbook of Crystal Growth* [1]. Readers are advised to refer to the extensive works of Scheel and colleagues [1,2] to get deeper insight into this incredible past.

Crystals have fascinated human civilization from prehistoric times, likely because of their beauty, clarity, and rarity. Watching the flatness of their faces, the sharpness of their edges, and the purity of their colors is a sheer delight. Along with this sense of delight, early civilizations must have wondered how nature was able to produce such a wide variety of crystals. This initial curiosity must have prompted them to try to prepare some crystals artificially, mainly by emulating nature, and the art (no science then) of growing crystals was born.

In ancient times, crystallization of salt from seawater and from salt solutions must have been practiced to get edible salt. Hence, crystal growth is an activity that is as old as human civilization. Generally, crystallization of salt was achieved by the evaporation of water from the solution, which in contemporary times is called the solvent evaporation method. This process, together with the fabrication of ceramics (practiced in Greece several centuries before Christ), must have been the oldest known technologies to transform materials into their crystalline forms. Subsequent development in this field must have occurred during the period of the great Roman Empire. The extended Roman Empire facilitated the transport of men and materials from places as far as Egypt and Asia Minor to places where finished products from metals, ceramics, and glasses were being mass produced (mostly Italy), and this must have increased the human knowledge of materials purification and preparation. Added to this, early knowledge developed in India and China in chemistry and metallurgy that passed through the Mediterranean region must have enriched this knowledge further. The alchemists of the time began to record their experiences on crystallization. For example, Pliny the Elder, who was an admiral in the Roman fleet, wrote about the purification and crystallization of copper sulfate (blue vitriol) and many other salts in the first century AD [1]. Much later, there were others like Geber (twelfth–thirteenth centuries) and Georg Bauer (sixteenth century) who also wrote on the crystallization process in their books [1]. In between, almost for thousand years, the activity in this field seems to be subdued, probably because of the rise of new religions (which were hostile toward rational thinking). Nevertheless, toward the end of the Middle Ages preparation of several salts was in pilot-plant scale. However, the basic understanding of the crystallization phenomena and crystal structure as such was quite vague.

The correlation between the external morphology and internal structure was first postulated by Kepler in 1611 [3]. He described the snowflake and from its regular hexagonal shape derived the hexagonal dense packing of water molecules. The concept of packing of spheres was further extended by Hooke [4] to crystal habits exhibited by crystals of common salt, alum, quartz, vitriol, and so on. This was followed by Nicolas Steno's extensive work on the origin of external forms of a variety of crystals [5]. Another important concept introduced by Steno was the reciprocity of crystal growth and crystal dissolution, which meant that those dissolved easily also grew easily.

Subsequently, in 1669, Steno [5] for the first time proposed the constancy of angles between the morphological planes of crystals, which became very useful in the identification of crystal habits. The need to measure the interfacial angles led to the development of goniometers, microscopes, and so on. Subsequent development in the field of mineralogy led to the introduction of symmetry elements, nonorthogonal crystal systems, and Miller indices.

With the advent of microscopes, the study of crystallization under microscope was possible, and Rouelle in 1745 used this technique to study the effect of supersaturation on growth stability [6]. Fifty years later, Lowitz [7], apart from describing many features of crystallization, also gave an account of seeded growth by which large crystals could be obtained.

In the nineteenth century, the crystal growth activity picked up, and extensive work on crystallization of bulk, epitaxial films, whiskers, and biological materials was reported. Mitscherlich [8] reported the melt growth of many known minerals. Crystallization from high-temperature solution was developed [9,10], allowing for a large number of oxide crystals to be grown. Also, by the end of the nineteenth century, solubility curves could be understood and metastable regions in the supersaturated solution could be measured. Gibbs [11] proposed the nucleation phenomena on thermodynamic grounds. Curie [12] tried to understand the equilibrium shapes of the crystals.

Until the beginning of the nineteenth century, apart from the usage of common salt in food, the only applications of crystals were in jewelry, medical prescription, and mysticism. For example [13]:

1. Kings in ancient India used crystals to protect themselves, in addition to using them in their crown jewelry.

2. Jade was recognized as a kidney healing stone in ancient China.

3. The ancient Chinese used crystals in feng shui, which is an art involving the placement of crystals in specific areas and in a particular orientation so that a person reaches harmony with his or her environment.

4. Ancient Egyptians used malachite and lapis lazuli with precious metals for healing and growth.

5. Early Greek and Roman writings refer to the use of stones as talismans.

6. It appears that Mayan and American Indians used crystals to diagnose and treat diseases.

7. Wearing crystals as amulets was one of the ancient healing practices in many nations.

8. In the Bible, it is recorded that Aaron wore a divinely inspired and extremely powerful "breastplate" containing 12 different crystals, which were studded on a silver plate in three rows of four crystals each. The 12 crystals used were sard, agate, chrysolite, garnet, amethyst, jasper, onyx, beryl, emerald, topaz, sapphire, and diamond.

9. Amethyst, which was very rare in the early centuries, was found in the crowns, rings, staffs, and jewelry of royalty and priests, and even in the Pope's ring.

10. Altar stones, candlesticks, and some crosses once carried seven main stones—diamond/clear quartz, sapphire, jasper, emerald, topaz, ruby, and amethyst.

11. In India, temple jewelry was studded with a variety of precious stones, with the most prominent among them being diamond.

However, technical use of crystals was yet to begin.

The twentieth century saw tremendous progress in science and technology, and with this, crystal growth technology also progressed rapidly. Many a time, overcoming the problems related to crystal growth became the rate-limiting step in the overall technological progress. Much of this progress was initially propelled by World War II, in which the potential of crystals in a variety of applications was exploited for military purposes. With the discovery of transistor action in semiconducting materials, the floodgates were opened in the field of electronics and microelectronics. Various other physical properties such as piezoelectricity, pyroelectricity, ferroelectricity, and other optoelectronic properties were exploited in sensor and detector applications. High hardness of materials such as carborundum and diamond was exploited in cutting and polishing. With the discovery of lasing action in ruby, crystals became amplifiers of coherent radiation.

A major input to the understanding of crystal growth phenomena came from the understanding of crystal structure, which was facilitated by x-ray diffraction, for which foundations were laid by Laue and colleagues [14,15] and Bragg [16,17]. As discussed in Chapter 3, the periodic nature of the crystal structure became the cornerstone in the development of crystal growth theories. With the knowledge of crystal structure it also became easy to predict various defect structures, which greatly influence a number of physical properties of the crystal.

To facilitate growth of crystals with diverse physicochemical properties, various crystal growth techniques were developed that were suited to obtain crystals from solid–solid, liquid–solid, and vapor–solid transitions. The industrial revolution that took place in the eighteenth and nineteenth centuries facilitated the development of sophisticated equipment for this purpose, which researchers could use to grow large and better crystals. With the progress in the field of electronics, precise control over growth parameters became possible. With this, growth of very high-quality crystals could be achieved, which further expanded the application potential of the crystals. Remote control of growth parameters became possible with automation. The crystal growth technology of today is hence far removed from what was

FIGURE 1.1
This cartoon was drawn by the author in his early days as a crystal grower.

being practiced in the early days (see Figure 1.1). We discuss some of these aspects in much greater detail in Section II of this book.

Organization of the Book

This book is intended for beginners. It is divided into two sections. The first section, "Principles," begins with this introductory chapter, which enumerates the historical development of the field of crystal growth and the motivation for the same. This is followed by a chapter each on nucleation, two-dimensional layer growth mechanism, defects in crystals, and screw dislocation theory of crystal growth with some later ideas. The section ends with a chapter on phase diagrams, which is especially important for this subject.

The second section, "Practice," deals with the experimental techniques of crystal growth. For a practicing crystal grower, this will be quite useful because it is here where the nuts and bolts of the subject are dealt with. Major techniques falling under solid–solid, liquid–solid, and vapor–solid equilibria have been discussed, along with personal anecdotes from the author as a crystal grower. Toward the end of the book, some of the characterization techniques that are essential to measure the quality of the grown crystals have been discussed. With characterization being such a vast field, the number of techniques discussed is not exhaustive, but adequate.

References

1. Scheel, H. J. 1993. Historical introduction. In *Handbook of crystal growth*, edited by D. T. J. Hurle, Vol. 1. Amsterdam: Elsevier, pp. 1–41.
2. Ellwel, D., and H. J. Scheel. 1975. *Crystal growth from high temperature solutions*. London: Academic Press.
3. Kepler, J. 1611. *Strenaseu de nivesexangula* [New year's gift concerning six cornered snow flake]. Frankfurt/Main: G Thampach.
4. Hooke, R. 1665. *Micrographia*. London.
5. Steno, N. 1669. De *solido intra solidum naturaliter contento* [Dissertation concerning a solid body naturally contained within a solid]. Florence: Jacobum Moukee.
6. Rouelle, G. F. 1745. Sur le selmarin (Premiere partie). De la crystallisation du Sel marin. In *Histoire de l'Acadêmie Royale des Sciences*, Paris, S57–S79.
7. Lowitz, J. T. 1795. Bemerkungenuber das Kristallisiren der Salze und Anzeige Einessichern Mittels, regelmassige Kristallenzuerhalten. *Chemische Annalen für die Freunde der Naturlehre, Arzneygelährtheit, Haushaltungskunst und Manufacturen* 1: 3–11.
8. Mitscherlich, E. 1922/23. Considering the materials which can crystallize in two different crystal forms. *Abhandlung Akad* 43 (see Mitscherlich's gesammelte Werke pp. 193).
9. Watt, G. 1804. Observations on basalt, and on the transition from the vitreous to the stony texture, which occurs in the gradual refrigeration of melted basalt; with some geological remarks. *Philosophical Transactions of the Royal Society* 94: 279–314.
10. Hall, J. 1805. Experiments on whinstone and lava. *Transactions of the Royal Society of Edinburgh* 5: 43–75.
11. Gibbs, J. W. 1878. *Collected works*. London: Longmans Green.
12. Curie, P. 1885. On training criteaux and capillary constants of their different faces. *Bulletin De La Societe Francaise De Mineralogie Et De Crystallographie*. 8: 145–150.
13. Inner Oracle. Crystals—Ancient light in sacred design. Available from: http://www.trinitysthomas.com/inner-oracle/crystals/
14. Friedrich, W., P. Knipping, and M. Laue. 1912. Interferenz-Erscheinungenbei Rontgenstrahlen [Interference phenomena with X-rays]. *Sitz. ber. Bayer. Akademie d. Wiss*. 303–322.
15. Laue, M. 1912. Eine Quantitative Prufung der Theorie fur die Interferenz-Ersch einungenbeiRontgenstrahlen. *Sitz. ber. Bayer. Akademie d. Wiss* 8: 363–373.
16. Bragg, W. L. 1912. X-rays and crystals. *Nature* 90: 219.
17. Bragg, W. L. 1912. The diffraction of short electromagnetic waves by a crystal. *Proceedings of the Cambridge Philosophical Society* 17: 43–57.

2

Nucleation Phenomena

Nucleation is one of the major mechanisms of first-order phase transitions that involve a latent heat. Here, a new phase emerges from the old phase whose free energy becomes higher than the generated new phase. Nucleation of the new phase, which is an extremely localized phenomenon, is the result of system fluctuations that bring a sufficient number of atoms or molecules together to form a stable size. Many natural processes happen due to nucleation, the most common among them being the formation of raindrops. Carbon dioxide bubbles nucleate shortly after the pressure is released from a bottle of carbonated liquid. Nucleation is also involved in such processes as cloud seeding and in instruments such as bubble chambers and cloud chambers. Of course, all natural and artificial crystallization processes start with a nucleation event. Most nucleation processes are physical, rather than chemical, although exceptions do exist (e.g., electrochemical nucleation).

Nucleation normally occurs at nucleation sites on surfaces contacting the liquid or vapor. Suspended particles or minute bubbles also provide nucleation sites. This is called heterogeneous nucleation. Nucleation without preferential nucleation sites is called homogeneous nucleation. Homogeneous nucleation occurs spontaneously and randomly. In this chapter, we deal with both homogeneous and heterogeneous nucleation processes.

Critical Supersaturation

It is a common experimental observation that during a phase transformation leading to crystal growth, the new phase will not appear when the system reaches the thermodynamical phase boundary (saturation) but only when it becomes supersaturated. In terms of the $P–T$ diagram shown in Figure 2.1, nucleation of a solid phase begins not at point 1 but at point 2. In the case of crystal growth from melt, the driving force would be the supercooling because, as can be seen from Figure 2.2, the free energy of the system is lower for the crystalline phase below the melting temperatures. However, above the melting temperature, it is the melt phase that has the lower free energy. The supersaturation ratio P_1/P_2 or the undercooling $\Delta T = T_1 - T_2$ corresponding to the onset of nucleation is called critical. The first written report on this phenomenon was given as early as 1724 by Fahrenheit [1], who describes

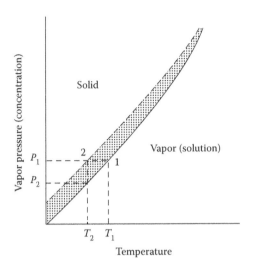

FIGURE 2.1
Vapor pressure and temperature phase diagram.

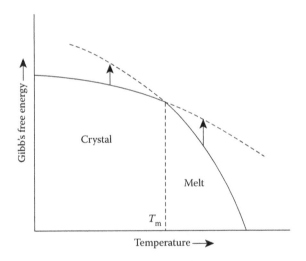

FIGURE 2.2
Dependence of Gibbs's free energy on T in the vicinity of melting point.

the abrupt solidification of supercooled water upon mechanical vibrations. In the following years, many data were gathered on the critical supersaturation or undercooling in inorganic salts crystallizing from aqueous solutions. It soon became apparent, however, that the degree of reproducibility of these data was very poor. The reason for this discrepancy is essentially due to the difficulty in deciding whether the observed nucleation happened on its own or with the help of an external agent.

The formation of a solid crystalline phase from liquid and gaseous solutions can occur only if some degree of supersaturation or supercooling has been achieved first in the system. The attainment of the supersaturated state is essential for any crystallization process, and the degree of supersaturation is the prime factor controlling the process. Any crystallization process can be considered to contain three basic steps:

1. Attainment of supersaturation or supercooling
2. Formation of critical nuclei
3. Growth of these nuclei into crystals

In practice, all three processes may be occurring simultaneously in different regions of the same crystallization vessel. The ideal crystallization, of course, would consist of a strictly controlled sequential procedure, but the complete cessation of nucleation cannot normally be guaranteed in a growth vessel where a suspended crystal is growing. The supersaturation of a system may be achieved by cooling, evaporation, the addition of a precipitant or diluting agent, or as a result of a chemical reaction between two homogeneous phases. However, the condition of supersaturation alone is not sufficient for crystal growth. Before crystals can grow, there must exist in the solution a number of minute solid bodies known as centers of crystallization, seeds, embryos, or nuclei. Nucleation may occur spontaneously (homogeneous), or it may be induced artificially (heterogeneous). It is not always possible, however, to decide whether a system has nucleated on its own or under the influence of some external stimulus.

Homogeneous (Spontaneous) Nucleation

Exactly how a crystal nucleus gets formed within a homogeneous fluid system cannot be described with any degree of certainty. To take a simple example, the condensation of a supersaturated vapor to the liquid phase is possible on a condensing surface only after the appearance of microscopic droplets called condensation nuclei. However, because the vapor pressure at the surface of these minute droplets is exceedingly high, they evaporate rapidly, even though the surrounding vapor is supersaturated. New nuclei form while old ones evaporate, until eventually stable droplets are formed by coagulation of a sufficient number of molecules under conditions of very high vapor pressure.

The formation of crystal nuclei is an even more difficult process. Not only do the constituent molecules have to come together, resisting the tendency to re-evaporate, but they also have to get positioned into a fixed lattice. The number of molecules in a stable crystal nucleus can vary from about

10 to several thousands. Water (ice) nuclei, for instance, contain about 80–100 molecules. The actual formation of such a nucleus can hardly result from the simultaneous attachment of the required number of molecules, which is an extremely unlikely event. It is likely that minute structures are formed, first from the collision of two molecules and then from that of a third with the pair, and so on. Short chains or flat monolayers may be formed initially, and eventually the lattice structure is built up. The construction process, which occurs very rapidly, can only continue in local regions of high supersaturation, and many of these "subnuclei" fail to achieve maturity. If, however, the nucleus grows beyond a certain critical size, it becomes stable under the average conditions of supersaturation in the bulk of the fluid.

The phenomenon of spontaneous nucleation can be understood on thermodynamic grounds by considering the various energy requirements. This was originally discussed by Gibbs [2]. When a group of freely moving molecules become aggregated into a solid state in which the molecular movement is much more restricted, a quantity of energy is released. For example, when a vapor condenses into a liquid, the latent heat associated with the change of state is liberated. In a given system, therefore, the transition from gaseous to liquid and then to a solid state represents a stepwise decrease in the degree of molecular mobility and likewise a decrease in the free energy of the system.

However, the formation of a liquid droplet or a solid particle within a homogeneous fluid demands the expenditure of a certain quantity of energy in the creation of the liquid or solid surface. Therefore, the total quantity of work W required to form a nucleus is equal to the sum of work required to form the surface, W_s (a +ve quantity) and the work required to form the bulk of the particle W_v (a −ve quantity). Hence,

$$W = W_s - W_v \tag{2.1}$$

For the formation of a spherical liquid droplet in a supersaturated vapor, Equation 2.1 becomes

$$W = a\sigma - \upsilon G_\upsilon \tag{2.2}$$

where G_υ is the energy gained per unit volume by the phase change and is independent of the radius and σ is the surface energy per unit area (surface tension).

Because the work done is equal to the energy change ΔG, we have for a spherical drop of radius r,

$$W = \Delta G = 4\pi r^2 \sigma - \frac{4}{3}\pi r^3 G_\upsilon \tag{2.3}$$

Now, the size and the energy of formation of a critical nucleus can be obtained by setting $(d\Delta G)/dr) = 0$ to locate the maximum, that is,

$$0 = 8\pi r\sigma - 4\pi r^2 G_\upsilon \quad \text{or} \quad G_\upsilon = \frac{8\pi r\sigma}{4\pi r} = \frac{2\sigma}{r}$$

Therefore,

$$\Delta G_c = 4\pi r^2\sigma - \frac{4}{3}\pi r^3 \frac{2\sigma}{r} = 4\pi\sigma r_c^2/3 \tag{2.4}$$

where we have replaced r by r_c.

Figure 2.3 is a plot of the three terms of Equation 2.3. The positive surface term rises as r^2, the negative volume term falls as r^3, and the resultant net energy change is in the form of a barrier with maximum ΔG at a certain critical radius r_c. This is the maximum thermodynamic energy barrier that separates the two phases. This graph clearly tells us that clusters of molecules that have radii less than the critical radius r_c will naturally tend to lower their energy by moving to the left of the curve, dispersing through evaporation (or melting), and sliding down the barrier to the vapor phase. Thus, all clusters with radii less than r_c are unstable because they find it energetically more favorable to disappear than to grow. Occasionally, a cluster will form with an initial radius r_c. Finding itself on the top of the energy barrier, it can go either to the left by evaporating and disappearing or to the right by growing and increasing its radius beyond r_c, thereby reducing the total energy. A condensate of radius r_c is thus the smallest nucleus capable of continued growth and hence is called a critical nucleus. For growth

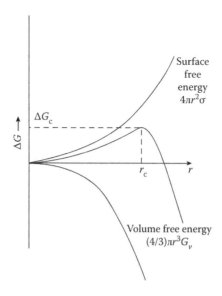

FIGURE 2.3
Change in free energy during the formation of a liquid droplet.

beyond r_c, the work required to create new interfacial area is increasingly offset by the energy gained from the transformation of volume to the lower energy crystalline phase. Below r_c, the energy loss by decreasing the interfacial area is more than that gained by the phase change; hence, the cluster dissolves.

Replacing r_c by $(2\sigma/G_v)$ in Equation 2.4, we have

$$\Delta G_c = 4\pi\sigma r_c^2/3 = 16\pi\sigma^3/3(G_v)^2 \tag{2.5}$$

The increase in the vapor pressures over a liquid droplet as its size decreases can be estimated from the Gibbs–Thomson formula [3], which may be written as

$$ln\left(P_r\big/p^*\right) = 2M\sigma/RT\rho r \tag{2.6}$$

where p_r and p^* are the vapor pressures over the liquid droplets of radius r and a flat liquid surface, respectively. In other words, p^* is the equilibrium saturation vapor pressure of the liquid, M is the molecular weight, ρ is the density of the liquid droplet, T is the absolute temperature, and R is the gas constant. Of course, it is easy to understand that p_r/p^* (equal to saturation ratio α) is nothing but a measure of supersaturation S (S being equated to $\alpha - 1$ by definition) so that we can write Equation 2.6 as

$$\ln \alpha = 2M\sigma/RT\rho r \quad \text{or} \quad r = \frac{2M\sigma}{RT\rho\ln\alpha} \tag{2.7}$$

If we substitute this value of r in Equation 2.4, we get

$$\Delta G_c = W = \frac{16\pi\sigma^3 M^2}{3(RT\,\rho\ln\alpha)^2} \tag{2.8}$$

The above equation is an extremely important relationship. It gives a measure of energy of nucleation in terms of degree of supersaturation of the system. It can be seen, for example, that when the system is only just saturated ($\alpha = 1$, $\ln \alpha = 0$), the amount of energy required for the formation of nucleus is infinite and so a saturated solution cannot nucleate spontaneously. It also suggests that any supersaturated solution can nucleate spontaneously because the work requirement associated with the process is finite. Of course, the quantity of energy required may be quite large, but the fact remains that spontaneous nucleation is theoretically possible at any degree of supersaturation.

Now, the rate of nucleation N (i.e., the number of nuclei formed for unit time per unit volume) can be expressed in the form of the Arrhenius equation.

$$N = A\ exp -(\Delta Gc/kT) \tag{2.9}$$

where A is the proportionality constant and called the frequency factor.

TABLE 2.1

Dependence of Nucleation Time on Saturation Ratio for Water

Supersaturation Ratio $\alpha = p/p^*$	1	2	3	4	5
Induction period	∞	10^{62} years	10^3 years	0.1 s	10^{-13} s
Remarks	No nucleation	Impractical	Impractical	Practical	Impractical

Now substituting the value of ΔG_c, we get

$$N = A\,exp\left[-\frac{16\pi\sigma^3 M^2}{3R^3 T^3 \rho^2 (\ln\alpha)^2}\right] \qquad (2.10)$$

This equation indicates that three main parameters govern the rate of nucleation: temperature T, degree of supersaturation α, and the interfacial tension σ. The dominant effect of the degree of supersaturation can be shown by calculating the time required for spontaneous appearance of nuclei in supercooled water vapor, which is given in Table 2.1 [4].

Therefore, in this case, a practical degree of supersaturation is in the region of $\alpha \sim 4.0$, but it is also clear that nucleation would occur at any value of α that is slightly greater than 1.0 if sufficient time is given. These predictions are in agreement with the experiments carried out on clean water vapor in cloud chambers used in nuclear studies.

As mentioned earlier, crystallization need not occur through spontaneous nucleation, and there are many ways of artificially inducing crystallization. However, excessive cooling does not aid nucleation. There is an optimum temperature for nucleation, and any nucleation below this value is unlikely. Agitation or bubbling of gas through the solution, mechanical shock, friction, and extreme pressures within melts and solution induce nucleation readily. Effects of external influences such as electric and magnetic fields, ultraviolet light, x-rays, β-rays, and sonic and ultrasonic radiations have often been studied. However, the best method for inducing crystallization is to seed the supersaturated solution with the small particles of material to be crystallized. The seeding process has been discussed in detail in Section II enumerating the techniques of crystal growth.

Heterogeneous Nucleation

Many reported cases of spontaneous nucleation are found on careful examination to have been induced in some way. Indeed, it is often suggested that true examples of spontaneous nucleation are rarely encountered. For example, a supercooled system can be seeded unknowingly by the presence

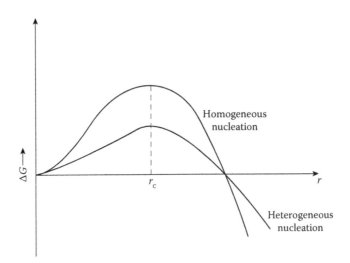

FIGURE 2.4
Difference in energy barriers for homogeneous and heterogeneous nucleation.

of atmospheric dust. Atmospheric dust often contains particles of the product itself, especially in industrial plants or in labs where samples of the crystalline materials are being handled. Because foreign particles are involved, this process is called heterogeneous nucleation. As the presence of the foreign body can induce nucleation at degrees of supersaturation lower than those required for a spontaneous nucleation, the overall free energy change associated with the formation of a critical nuclei under heterogeneous conditions ($\Delta G_c'$) must be less than the corresponding free energy change (ΔG_c) associated with homogeneous nucleation. Therefore,

$$\Delta G_c' = \phi \Delta G_c \quad \text{where } \phi < 1 \tag{2.11}$$

Because of this, the barrier height for heterogeneous nucleation is reduced and consequently less supersaturation/supercooling is needed. As is shown in Figure 2.4, the critical radius remains unchanged. However, the volume can be significantly less for heterogeneous nucleation due to wetting, which will change the shape of the nucleus.

Nucleation on a Substrate

It is an established fact that substrates promote the formation of nuclei. If the substrate is of the same material (as is the case with seeded growth), nucleation occurs preferentially over the substrate surface rather than in

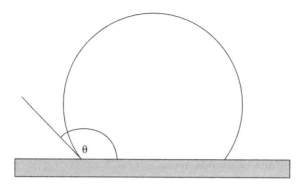

FIGURE 2.5
Equilibrium shape of a liquid drop on a flat solid substrate.

a homogeneous phase. When the nucleating phase is isotropic, its shape can be approximated to a portion of a sphere. If the substrate is perfectly flat, then the shape of the nuclei can be defined by the angle of contact θ, which is defined as the angle between the tangent to the sphere at the contact point and the substrate (Figure 2.5). One can show, then, that the change in the free energy ΔG_{sub} (or work of formation) for the formation of critical nucleus on the substrate as given by Becker and Doring [5] is

$$\Delta G_{c\text{-}sub} = \Delta G_c\{(1/2) - (3/4)\cos\theta + (1/4)\cos^3\theta\} \qquad (2.12)$$

where ΔG_c corresponds to work of formation of a critical nucleus in the homogeneous phase at the same supersaturation. As can be seen from the above expression, $\Delta G_{c\text{-}sub}$ varies from ΔG_c to zero when θ changes from 180° to 0°. In the first case, the substrate has no influence on the nucleation process because there is no wetting of the substrate. However, when $\theta = 0$, there is complete wetting and the nucleation becomes essentially two dimensional.

Nucleation of a Crystalline Material

The formation of solid nuclei from a parent phase can be considered analogous to the condensation of water droplets, even though a liquid droplet is very different from a crystalline particle. However, some changes have to be made in the theory. First, the interfacial energy must be the one appropriate to the boundary between the higher energy parent phase and the solid, rather than a liquid. Second, a different measure for the deviation from the equilibrium is required. In nucleating crystals from solution, pressure ratios are not suitable; instead, a ratio of dissolved solution concentrations is used (c/c^*). In nucleating crystals from a melt, a ratio of temperatures is used. Finally, the frequency factor A is bound to be larger, if the parent phase is a dense liquid. Because there are many more atoms per unit volume of the liquid than in a vapor, chances of nucleation would be higher. These changes are of minor importance and the essential behavior remains the same; that is, at a critical departure from equilibrium, a homogeneous system will become critical and nucleate crystallites, which will subsequently grow. Realizing this analogy, certain assumptions made by Gibbs regarding the thermodynamics of nucleation of a liquid phase were extended to the nucleation of crystalline phase from vapor, liquid, and also from other solid phases, though they were strictly not applicable. The classical nucleation theory thus developed [6] for example, incorporated an approximation known as capillarity approximation, which assumes that the molecular arrangement in the crystal nuclei is identical to that of the large crystal. This would mean that the surface free energy of the nuclei is equal to that of the crystal interface, which is not true in all cases, and this led to the discrepancies between the theoretical and experimental data on nucleation in certain cases. Nevertheless, the classical nucleation theory developed by extending the approach of Gibbs was highly successful in predicting homogeneous nucleation rates for many materials [7–9].

Equilibrium Shape of Anisotropic Nuclei

In the sections discussed earlier, the interfacial energy (surface tension) of the nucleating phase was treated as orientation independent; that is, the phase was considered to be isotropic, which is, of course, true for liquids. When one deals with the anisotropic phase, which is the case with crystals, we should also address the following:

1. The shape of the critical nucleus for which the specific surface free energy is minimum
2. The combination of specific surface free energies that results in the equilibrium shape of small crystals

Therefore, the main problem concerning the nucleation of anisotropic phase is that of minimization of its total surface free energy for a given size, taking specific surface free energies of individual developing faces into account, that is, its equilibrium shape. Wulff's [10] treatment on equilibrium shape of a growing crystal has been discussed in Chapter 3.

References

1. Fahrenheit, D. B. 1724. Experimentaet Observationes de Congelationeaquae in vacuofactae. *Philosophical Transactions of the Royal Society* 33: 381–391.
2. Gibbs, J. W. 1876. On the equilibrium of heterogeneous substances. *Transactions of the Connecticut Academy of Arts and Sciences* 3: 108–248, 343. (Also see *Scientific papers* Vol I. 1906. London: Longmans Green).
3. Mullin, J. W. 1961. *Crystallization*. London: Butterworths, p. 104.
4. Volmer, M. 1939. *Kinetic der Phasenbildung* (Kinetics of the phase formation). Dresden, Germany: Steinkopff.
5. Becker, R., and W. Doring. 1935. Kinetische Behandlung der Keimbildung in übersättigten Dämpfen. *Annalen der Physik (Leipzig)* 24: 719–752.
6. Katz, J. L., and B. S. Ostermier. 1967. Diffusion cloud-chamber investigation of homogeneous nucleation. *Journal of Chemical Physics* 47: 478–487.
7. Katz, J. L., C. J. Scoppa, II, N. G. Kuman, and P. Mirabel. 1975. Condensation of a supersaturated vapor. II. The homogeneous nucleation of the n-alkyl benzenes. *Journal of Chemical Physics* 62: 448–465.
8. Katz, J. L., P. Mirabel, C. J. Scoppa, II, P. Mirabel, and T. L. Virkler. 1976. Condensation of a supersaturated vapor. III. The homogeneous nucleation of CCl_4, $CHCl_3$, CCl_3F, and $C_2H_2Cl_4$. *Journal of Chemical Physics* 65: 382–392.
9. Becker, C., H. Reiss, and R. H. Heist. 1978. Estimation of thermophysical properties of a large polar molecule and application to homogeneous nucleation of l-menthol. *Journal of Chemical Physics* 68: 3585–3594.
10. Wulff, G. 1901. Zur Frage der Geschwindigkeit des Wachsthums und der Auflösung der Krystallflachen. *Zeitschrift fur Kristallographie* 34: 449–530.

3

Crystal Growth Mechanisms

Even though x-ray diffraction techniques have enabled scientists to reveal the internal structure of crystals, they did not contribute much to the understanding of the very outer layers of atoms in the crystal. It will be seen later that the surface layers play a very important role in the growth of crystals because they act as collectors of fresh material. In the early part of the twentieth century, a number of techniques such as electron microscopy, phase-contrast microscopy, and multiple-beam interferometry were developed to study the surface of crystals. The knowledge gained by such studies has provided valuable information concerning the way in which crystals grow. Some very striking and precise observations made about the crystal surfaces have put the theory of crystal growth on sound footing. In this chapter, we look at some of the theoretical frameworks of crystal growth, which are useful for understanding what actually happens when a crystal is growing. With the help of these theories, we can infer the influence of various growth parameters on the crystallization process.

In the previous chapter, we discussed the nucleation of crystals. As soon as stable nuclei (i.e., particles larger than the critical size) are formed in a supersaturated solution, they begin to grow into crystals of visible size with predetermined flat facets. It has been found that even when we start the growth on a spherical seed, the ultimate shape of the crystal is polyhedral bounded by flat faces that grow with the lowest rates [1]. Why should the crystal faces be flat? The rates of growth of crystal faces are in general a function of supersaturation of the fluid phase from which they grow. If so, why should the habit faces of the crystal grow truly planar when in many cases, the conditions of supersaturation may vary from point to point on these faces? Many attempts have been made to explain the mechanisms and rate of crystal growth, and in this chapter, we briefly review the most prominent among them.

Early Theories

The first quantitative theory of a crystal growth was given by Gibbs himself on thermodynamic grounds. Essential features of his theory are summarized here. An isolated droplet of a fluid is most stable when its surface free

energy and hence its area are a minimum. Gibbs [2] suggested that
the growth of a crystal could be considered as a special case of this principle: The total free energy of a crystal in equilibrium with its surroundings
at constant temperature and pressure would be a minimum for a given volume. Taking the volume free energy per unit volume as constant throughout the crystal, the condition becomes

$$\sum_{i=1}^{n} \sigma_i\, F_i = \text{minimum} \tag{3.1}$$

where σ_i is the surface free energy per unit area of the ith face of area F_i on a
crystal bounded by n faces. Thus, only those faces will develop that lead to a
minimum total surface free energy for a given volume.

Of course, a liquid droplet is very different from a crystalline particle.
In the former the constituent atoms are randomly dispersed, whereas in the
latter they are regularly arranged on a lattice. Gibbs was fully aware of the
limitations of his simple analogy, but Curie [3] took up this idea as a useful
starting point to develop a general theory of crystal growth. He calculated
the shapes and final forms of crystals in equilibrium with solution/vapor
consistent with the condition that the free energy is a minimum for a given
volume. Subsequently, Wulff [4] extended this theory further. He suggested
that the equilibrium shape of a crystal is related to the free energies of the
faces and showed that the crystal faces would grow at rates proportional
to their respective surface free energies. His treatment of the problem is as
follows.

Considering a point P within the crystal (Figure 3.1), let p_i be perpendicular from P on the ith face of area F_i. Then, the total volume V of the crystal
will be equal to the sum of the volumes of pyramids with P as vortex and
the faces as bases.

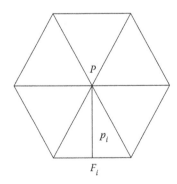

FIGURE 3.1
Wulff's treatment of the equilibrium shape of a crystal.

Therefore,

$$V = \frac{1}{3}\sum_{i=1}^{n} p_i F_i \tag{3.2}$$

and the total free energy E is given by

$$E = \sum_{i=1}^{n} \sigma_i F_i \tag{3.3}$$

Now, if one set of faces grows at the expense of another set, the change in volume dV may be written as

$$dV = \sum_{i=1}^{n} F_i \, dp_i \tag{3.4}$$

Differentiating Equation 3.2, we get

$$dV = \frac{1}{3}\sum_{i=1}^{n} (F_i dp_i + p_i dF_i) \tag{3.5}$$

If we keep the volume constant, $dV = 0$. Then, we have from Equations 3.4 and 3.5

$$\sum_{i=1}^{n} p_i \, dF_i = 0 \tag{3.6}$$

Now, if the total free energy is constant and if we neglect any variation of σ with respect to faces F, we have from Equation 3.1

$$\sum_{i=1}^{n} \sigma_i \, dF_i = 0 \tag{3.7}$$

Hence, from Equations 3.6 and 3.7, we get

$$p_i \; \alpha \; \sigma_i \tag{3.8}$$

or

$$\sigma_i / p_i = \text{constant} \tag{3.9}$$

Therefore, the crystal should form a polyhedron such that the "perpendicular" distances from a point P within the crystal are proportional to specific free energies of the appropriate faces. Wulff had also indicated that the rate of growth of a face would be inversely proportional to the reticular density of the respective lattice plane so that the faces having low reticular density would rapidly grow and eventually disappear. Hence, the faces exhibited by the crystal would be usually of the low-index planes.

Further modification of surface energy theory was due to Marc and Ritzel [5]. They considered that

1. Under the influence of surface tension alone, the crystal is constrained to adopt a form consistent with $\sum_{i=1}^{n} \sigma_i F_i = \text{minimum}$.

2. Since the dissolution rate has different values in different directions (measured by the solubility of isolated faces), the crystal generally favors the more soluble faces at the expense of the less soluble ones (with low soluble ones vanishing) in attaining the equilibrium shape.

Hence, the crystal will assume an equilibrium shape that takes into account both of these tendencies. When the differences in solubility are very small, the first of these will dominate, resulting in surface energy dominating the growth form (as is required by the Gibbs relation).

Wulff's idea of equilibrium shape of the growing nucleus was extended to nucleation on a substrate by Kaischew [6] half a century later, by which time basic theories of crystal growth had been reasonably developed. This theorem, which has been referred to as the Wulff–Kaischew theorem, can be written as

$$\sigma_n/p_n = (\sigma_i - \beta_i)/p_i = \text{constant} \tag{3.10}$$

where σ_i and p_i are as already defined and β is the free energy of adhesion. Adhesion free energy is intimately associated with wetting of the substrate by the crystal formed on it because, as we can infer from Equation 3.10, when $\beta = 0$, the distance p_n will have its value corresponding to the case when there is no substrate, which implies no wetting. At the other extreme, when β is equal to or larger than the adhesion energy ($\sigma_i + \sigma_s = 2\sigma$), we have complete wetting and the three-dimensional crystal is reduced to an island of monolayer thickness. In the intermediate cases, we have incomplete wetting and the crystal height p_i will be in between the two extremes.

Two-Dimensional Layer Growth Mechanism

However, there is no general acceptance of the early theories of crystal growth because there is little quantitative evidence available to support them. The reason for their nonacceptance is apparent. Even though these scientists were fully aware of the basic difference between a liquid droplet and the crystalline particle, they never made an attempt to incorporate the periodic nature of the crystalline materials in their theories. This was first done by Kossel [7] in the year 1927. Kossel analyzed the atomic in-homogeneity of a crystal surface and explained the fundamental importance of molecular steps and molecular kinks along the steps to the growth process. Based on the Kossel model, Stranski [8], Stranski and Kaischew [9], Volmer [10], and others [11] built up a conventional theory of crystal growth.

The basic tenets of this theory can be illustrated with the aid of Figure 3.2. This figure illustrates the features on the surface of a growing crystal wherein the crystal is assumed to possess a simple cubic lattice, the bonds are homo-polar and nondirectional (e.g., van der Waals); and the face considered is a (100) face. Each cube in Figure 3.2a represents one molecule in its rest position.

Let W be the binding energy of a molecule in the crystal, with the binding energy in the gas phase being taken as zero. This binding energy is equal to the work done against the intermolecular forces to take out the molecule from its particular position in the lattice to infinity. This will be a +ve quantity that increases with increasing stability of the molecule in the lattice. In general, the total binding energy of a molecule in the lattice is equal to the sum of the binding energies that would exist between the molecule and each of the other molecules in the lattice taken individually at their appropriate distances.

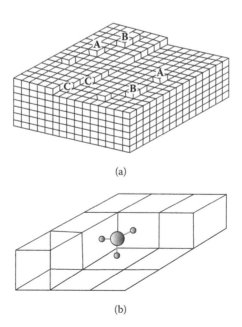

(a)

(b)

FIGURE 3.2
(a) Growth on a crystal surface by the attachment of an atom/molecule at the kink site and (b) magnified view of a kink site.

Naturally, only molecules in close proximity will contribute significantly to the total binding energy. In the present discussion, we neglect all molecular interactions beyond the nearest neighbors. Let the binding energy between a pair of nearest neighbors be ϕ. On this basis, let us consider the binding energy of one of the molecules labeled A in Figure 3.2. Such a molecule is called a face-adsorbed molecule. This molecule has only one nearest neighbor, which is directly beneath it; hence, its binding energy will be equal to ϕ. Likewise, the molecule B will have the binding energy 2ϕ (because such molecules have two nearest neighbors: one below and the other on the side), and those labeled C will have 3ϕ. The position of the molecule labeled C is called the repeatable step (kink site). It is obvious that by adding molecules at the repeatable step, we can complete a line of molecules. While starting another line, the first arriving molecule will have a binding energy 2ϕ, but once the new line has started, the binding energies will be 3ϕ again for the rest of the line. Hence, except in very small crystals, the energy of the complete crystal will be approximately equal to the energy of the repeatable step multiplied by the number of molecules in the crystal.

For the three types of molecules considered earlier (Figure 3.2), the binding energies are

1. Face-adsorbed molecules labeled A: ϕ
2. Molecules labeled B: 2ϕ
3. Molecules at repeatable step or kink site labeled C: 3ϕ

Due to the very low binding energy at the face-adsorbed positions, the molecules will have a great tendency to re evaporate. The much greater binding energy of the molecules at the repeatable step would suggest that any molecule sitting at this site would be strongly held and would tend to stay. The growth of crystals was therefore considered to occur by the addition of molecules at the repeatable step. Molecules coming from the fluid phase would arrive randomly at various surface sites, but those arriving at the repeatable step would stay and the rest of the molecules would re evaporate.

The growth of crystal by the attachment of the molecules at the kink site would then occur in the following way. When a row of molecules is completed, there would be no further repeatable step available; hence, the next stage would be the starting of a new row. Since the molecule starting the next row has a binding energy of only 2ϕ, the lifetime of such a molecule on the surface would be short. However, rather rarely two molecules may be adsorbed simultaneously on adjacent sites at the edge of a completed row, and if this occurs, both molecules by this very event become repeatable step molecules, each with energy 3ϕ, and a new row will then be able to grow. As a result, there will be a delay at the end of every row before the next one starts. Likewise, when the last row is completed, resulting in an edgeless face, the starting of the next layer is expected to give rise to an even longer delay. To start a new layer, it would be necessary for a group of molecules to be adsorbed in adjacent sites so that they could stabilize each other sufficiently for further growth to proceed by the repeatable step mechanism. The group of molecules needed to start a new layer is referred to as the surface nucleus. Furthermore, we can assume that the time required for the formation of the surface nucleus would determine the rate of growth.

The earlier argument has been based on a model in which the molecule is assumed to be held by the nearest neighbor force only. The same sort of conclusion will emerge, however, even if more distant interactions are taken into account. An important consequence of the repeatable step mechanism is that a crystal appears to grow by the spreading of monomolecular layers with straight edges. Consequently, crystal surfaces would be truly flat.

Now, we can see that if the growth were to occur as stated earlier, the numerical results that the mechanism would give for the growth rate must be very much less than the experimental values. This is because the basic postulate of the theory requires that only those molecules that hit the surface at the repeatable step would condense and join the lattice and those that arrive elsewhere at the face will re evaporate, since they will have much lower binding energy. Since the area of the repeatable step site is only an immensely small fraction of the area of the face of the crystal of visible dimensions, the fraction of the colliding molecules that contributes to the growth of the crystal will be very small. Furthermore, the rate of formation of a new layer through surface nucleation would also be vanishingly small, and this would reduce the overall rate of growth of the face still further.

Volmer [10] in the year 1939 noted one obvious shortcoming in the aforementioned model of growth. While studying the growth of the mercury crystals from vapor, he realized that the face-adsorbed molecules diffuse with considerable ease over the crystal surface before reevaporating. He argued that if the face-adsorbed molecule can migrate over the surface, this would lead to a much greater rate of condensation, since the migrating molecules would be able to reach the growing step and get attached. From simple calculation (Appendix I), one can show that the mean diffusion distance x_s of the face-adsorbed molecule is given by

$$x_s = ae^{(W_s' - U_s)/2kT} \qquad (3.11)$$

where U_s is the activation energy between two adjacent equilibrium sites on the surface at a distance a from each other and W_s' is the energy of evaporation from the surface to the vapor. On a (111) surface of a face-centered cubic (fcc) crystal, x_s could be as high as $\sim 4 \times 10^2 a$.

Further, Frenkel [12] pointed out that a mono molecular step on a crystal surface contains a high concentration of kinks. Burton and Cabrera [13] calculated (Appendix II) the concentration of kinks in a step and expressed it as

$$x_0 \sim \frac{1}{2} a \, exp(w/kT) \qquad (3.12)$$

where x_0 is the mean distance between the kinks in a step and w is the energy necessary to form a kink. When this formula is applied to a close-packed step on a (111) face of an fcc crystal, one gets a value of $4a$ for x_0, which means we shall have a kink for every four sites in the step. This result is very important from the point of view of the rate of advance of the growing steps.

Due to accretion of the molecules at the kink sites, a straight step will advance. Since the rate of arrival of the molecules depends on vapor pressure, when the vapor pressure is increased above the equilibrium value, more molecules join kinks than leave them and hence the step advances. From these considerations, one can derive a formula for rate of advance of the step υ_∞. This is given in Appendix III and the result is

$$\upsilon_\infty = 2(\alpha - 1)\, x_s\, \upsilon\, e^{-W/kT} \tag{3.13}$$

for a straight step. Here, α is the supersaturation ratio and υ is the frequency factor (for monoatomic crystals $\upsilon \sim 10^{13}$ s^{-1}).

For a curved step with radius of curvature ρ, the velocity υ_ρ is given by

$$\upsilon_\rho = \upsilon_\infty\, [1 - (\rho_c/\rho)] \tag{3.14}$$

The equations for the rate show that the velocity with which the edge of the layer travels is proportional to supersaturation ($S = \alpha - 1 = \Delta p/p_0$). This would mean that steps initially present on a crystal surface will travel with a velocity proportional to supersaturation toward the edge and disappear, resulting in a flat surface. Such a surface will probably contain a few face-adsorbed molecules and perhaps a few vacancies. Now any further growth will occur only if new steps are formed.

Surface Nucleation

How can these new steps be formed on perfect crystal faces? One may think that such steps could be created by thermodynamic fluctuations in much the same way as the kinks are formed at the step. However, this does not happen unless the temperatures exceed a certain critical value that may be even higher than the melting point. We should realize that for the growth to continue, steps must be formed gradually at growth temperatures. As mentioned earlier, for creating a new step, a two-dimensional nucleus or island monolayer of certain critical diameter has to be formed on the edges of which growth can proceed (Figure 3.3). Once such a layer is formed, it spreads rapidly at a rate proportional to the imposed supersaturation across the whole face, and the process of growth is held up again until a new nucleus is formed on that surface. The probability that a nucleus is formed at any point per unit time

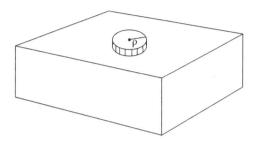

FIGURE 3.3
Surface with an island of monolayer thickness.

depends on the degree of supersaturation at that point, but the mechanism as such ensures that the rate of growth of the whole surface is uniform.

The problem of nucleating an island monolayer on a crystal surface is similar to the familiar problem of nucleating a water droplet from vapor, except for the dimensionality. For a given degree of supersaturation, there is a critical radius ρ_c of the surface nucleus (assuming it to be circular) such that the nucleus of radius greater than ρ_c will grow and if less will evaporate. This arises because the line energy of the step of the surface nucleus creates a local equilibrium vapor pressure that is inversely proportional to its radius of curvature. Thus, a surface nucleus with a small radius of curvature will have a high equilibrium vapor pressure, which may exceed the ambient vapor pressure, resulting in its evaporation.

An elementary estimate of rate of nucleation, mainly due to Mott [14], is as follows. Assuming the nucleus to be circular of radius ρ, its free energy F is given by

$$F = -\pi\rho^2\omega + 2\pi\rho\sigma \quad (3.15)$$

where ω is the free energy for unit area gained when molecules condense from vapor onto the surface and σ is the free energy per unit length of the step at the boundary of the nucleus. From this expression, it is shown that F increases at first as ρ increases from zero, reaches a maximum value, and thereafter decreases with increase in ρ, as shown in Figure 3.4.

The position of this maximum can again be found by differentiating the equation with respect to the radius and equating it to zero.

$$dF/d\rho = -2\pi\rho\omega + 2\pi\sigma = 0 \quad (3.16)$$

where

$$\rho = \rho_c = \sigma/\omega$$

Therefore,

$$F_c = \pi\sigma^2/\omega \quad (3.17)$$

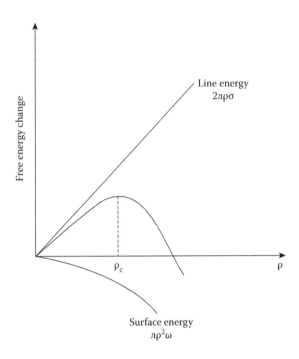

FIGURE 3.4
Free energy change during surface nucleation of an island monolayer.

An island of molecule smaller than radius ρ_c has a higher probability of evaporating than growing, while those larger than ρ_c have a higher probability of growing.

Now, the probability Γ that such a nucleus is formed in unit time and unit area can be written as

$$\Gamma = ne^{-F_c/kT} = n^{-\pi\sigma^2/\omega kT} \tag{3.18}$$

where n is the number of molecules of the vapor arriving at unit area per unit time. From detailed calculations, we can show that

$$\omega = kT \ln \alpha/a^2 \tag{3.19}$$

where α is the saturation ratio and therefore

$$\rho_c = \sigma/\omega = a\varphi/2kT\ln \alpha \tag{3.20}$$

where σa has been replaced by $(\tfrac{1}{2})\varphi$, half the bond energy.

Hence, the probability per unit time per unit area that an island nucleus is formed is given by

$$\Gamma = ne^{-F_c/kT} = ne^{-\pi a^2\sigma^2/k^2T^2\ln \alpha} = ne^{-\pi\varphi^2/4k^2T^2\ln \alpha} \tag{3.21}$$

However, Burton et al. [15] have reexamined this problem with a more detailed consideration of the shape of the nucleus. Their rigorous analysis shows that the nucleus is not really circular and that σ the free energy per unit length of the edge varies with crystallographic direction. In the more general case, they considered ρ_c to be the radius of the circumscribing circle to the critical nucleus. This essentially led to the substitution of a more realistic pre-exponential factor in Equation 3.21, and the rate of the formation of the nuclei is given by

$$\acute{\Gamma} = Z(A/s_0)\, e^{-F_c/kT} \tag{3.22}$$

where Z is the rate of arrival of molecules at single surface lattice site, A is the surface area of the crystal under consideration, and s_0 is the area per molecule in the layer.

Verification

Volmer and Shultz [16] experimentally measured the growth rate of iodine, naphthalene, and white phosphorous crystals from their vapors at 0°C as a function of supersaturation under low supersaturation conditions. Their experimental values show that growth rate is proportional to $(\alpha - 1)$ down to the lowest attainable value (0.01), except in the case of iodine for which $(\alpha - 1)$ was less than 0.02. In this case, the growth rate was proportional to $(\alpha - 1)^2$.

Now let us see what Equation 3.22 gives for 2% supersaturation. Kinetic theory of gases shows that even at appreciable vapor pressure, Z cannot be greater than $10^{13}\,s^{-1}$. For a crystal of millimeter dimensions, A/s_0 will not be greater than 10^{14} and will generally be more like 10^9. (Note that I_2 crystallizes in an orthorhombic lattice with the lattice parameters $a = 4.795$ Å, $b = 7.225$ Å, and $c = 9.780$ Å.) The value of ϕ/kT for I_2 at 0°C is about 6. Taking the upper limit for A/s_0 and substituting other values, we get

$$\acute{\Gamma} = 10^{27}\, e^{-\Pi \cdot \frac{36}{4} \cdot \frac{1}{\log 1.02}} = 10^{27}\, e^{-\Pi \cdot 9 \frac{1}{0.0086 \times 2.303}}$$

$$\approx 10^{27}\, e^{-2800} \approx 10^{27} \times 10^{-1200}$$

Hence, it is seen that the rate of formation of a surface nucleus contains 10^{-1200} as a factor. This is so extremely small that the calculated rate of growth will be effectively zero for any reasonable value of pre-experimental factor. In fact, detailed calculation using reasonable values ϕ/kT shows that for an observable growth rate (say, 1 μ/month), we need a supersaturation of about 25%–50%. The conventional theory could only conclude that growth of crystals could not take place by a surface nucleation mechanism at lower supersaturation. A new mechanism had to be invented for growth of crystals under low supersaturation. At this juncture, Frank came to the rescue of the conventional theory by drawing the attention of probable role of defects, in particular, of screw dislocations in furthering the growth process. This led to the development of screw dislocation theory of crystal growth. Before discussing this theory, we will digress a little and have a brief introduction to crystal defects. This will be dealt with in Chapter 4. Such an introduction will also be useful when we take up "crystal characterization."

References

1. Spangenberg, K. 1934. Wachstum und Auflosung der Kristalle. In *Handwort-erbuch der Naturwissenschaften*, edited by G. Fischer, Vol. 10, 2nd ed. Jena, Germany: Gustav Fischer, pp. 362–401.
2. Gibbs, J. W. 1961. *The scientific papers*, Vol. 1. New York, NY: Dover, p. 55.
3. Curie, P. 1885. On training criteaux and capillary constants of their different faces. *Bulletin De La Societe Francaise De Mineralogie Et De Crystallographie* 8: 145–150.
4. Wulff, G. 1901. Zur Frage der Geschwindigkeit des Wachstums und der Auflösung von Krystallflächen [On the question of the speed of growth and the desolution of crystal faces]. *Zeitschrift für Kristallographie* 34: 449–530.
5. Marc, R., and A. Z. Ritzel. 1911. Ueber die Faktoren, die den Kristallhabitus be-dingen. *Zeitschrift für Physikalische Chemie* 76: 584–590.
6. Kaischew, R. 1951. Equilibrium shape and work of formation of crystalline nuclei on substrates. *Bulgarian Academy of Sciences (Phys)* 1: 100–133.
7. Kossel, W. 1927. Zur theorie des krystallwachstums. *Nachrichten von der Gesellschaft der Wissenschaften zu Göttingen, Mathematisch-Physikalische Klasse* 135–143.
8. Stranski, I. N. 1928. The theory of crystal growth. *Zeitschrift fur Physikalische Chemie* 136: 259–278.
9. Stranski, I. N., and R. Kaischew. 1934. The kinetic description of the nucleation rate. *Zeitschrift fur Physikalische Chemie (B)* 26: 317–326.
10. Volmer, M. 1939. *Kinetic der Phasenbildung* [Kinetics of the phase formation]. Dresden, Germany: Steinkopff.
11. Becker, R., and W. Doring. 1935. Kinetische behandlung der keimbildung über sättigten dampfen. *Annalen der Physik (Leipzig)*. 24: 719–752.
12. Frenkel, J. 1945. Viscous flow of crystalline bodies under the action of surface tension. *Journal of Physics USSR* 9: 385–391.
13. Burton, W. K., and N. Cabrera. 1949. Crystal growth and surface structure. Part II. *Discussions of the Faraday Society* 33: 40–48.
14. Mott, N. F. 1950. Theory of crystal growth. *Nature* 165: 295–297.
15. Burton, W. K., N. Cabrera, and F. C. Frank. 1951. The growth of crystals and the equilibrium structure of their surfaces. *Philosphical Transactions of the Royal Society A* 243: 299–358.
16. Volmer, M., and W. Schultze. 1931. Condensation on crystals. *Zeitschrift fur Physikalische Chemie* 156: 1–22.

4

Defects in Crystals

An ideal crystal in which every lattice site is occupied by an atom, a molecule, or a group of atoms is a geometric concept. In real crystals, a variety of deviations from the ideal lattice structure occur. Any deviation from the perfect atomic arrangement in a crystal is considered an imperfection or a defect. It has now become clear that many important properties of solids are controlled as much by imperfections as by the nature of the host crystal, which may act only as a solvent or matrix for the imperfections. As we know, the electrical conductivity of doped semiconductors is entirely due to a trace amount of intentionally added impurities. Considerable deviations from the theoretical values of the measured intensities of x-rays diffracted from crystals occur due to the presence of defects. The colors of many crystals are due to imperfections. Luminescence in crystals is nearly always connected to the presence of impurities. Diffusion of atoms through solids may be accelerated enormously by impurities and imperfections, and finally the mechanical properties of solids are usually controlled by imperfections.

Imperfections or defects in the crystalline lattice are of different kinds. They are usually classified according to their geometrical nature as point, line, surface, and volume defects. As the names themselves indicate, a point defect has the center of disruption at a point in the structure, line imperfection causes disruption along a line, surface defect is found along a surface, and volume defect is manifested in the bulk. Following are the commonly observed defects in crystals under each category (Table 4.1).

In addition to these, one can also include excitons and polarons to the list under point defects. However, because of their special nature, these are not discussed here. In much of the earlier work, the presence of different kinds of defects was inferred in order to account for certain crystalline properties, but continuous improvements in the resolution of the electron microscope have confirmed many of these by more direct methods.

Point Defects

Point defects in a crystal can be grouped into two principal categories: native and foreign defects. A so-called point defect may be complex if it includes several kinds of foreign atoms in close proximity or foreign atoms closely

TABLE 4.1

Crystal Defects

Point Defects	Line Defects	Surface Defects	Volume Defects
Vacancy	Dislocations	Grain boundaries	Voids
Schottky	(edge, screw, mixed, and partial)	Low-angle grain boundaries	Inclusions
Interstitial		Twin boundaries	
Frenkel		Stacking faults	
Substitutional			
Color centers			

associated with some native defect. This sounds like a contradiction to the term "point" defect; however, the maximum dimension of these point defects in any direction is no more than a few atomic spacing.

Vacancy

The simplest among the native defects is a vacancy (Figure 4.1), a lattice site that should have an atom but does not. Since the regular nature of the crystal results from a delicate balance of forces at each equivalent lattice site, we might expect that the absence of an atom would change the balance of forces near the vacancy. In fact, the atoms in its neighborhood will shift their positions slightly inward until a new equilibrium configuration is reached. Far away from the vacancy, the effect of absence will be negligible so that the distortion satisfies our condition of being confined mostly to a small, roughly spherical region around it.

The equilibrium concentration of vacancy can be estimated in a simple way as follows.

The probability that a given site has a vacancy is proportional to the Boltzmann factor

$$P = exp\,(-E_v/kT) \tag{4.1}$$

where E_v is the energy required to form a vacancy.

If there are now n_0 atoms in the crystal, the number of vacancies n at a given temperature T is given by

$$\frac{n}{n_0 - n} = e^{-E_V/kT} \tag{4.2}$$

If $n_0 >> n$, we can write

$$\frac{n}{n_0} = e^{-E_V/kT} \qquad \therefore n = n_0 e^{-E_V/kT} \tag{4.3}$$

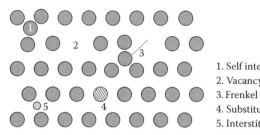

1. Self interstitial atom
2. Vacancy
3. Frenkel
4. Substitutional impurity atom
5. Interstitial impurity atom

FIGURE 4.1
Different types of point defects.

If $E_v = 1$ eV, and $T = 1000$ K, then $n/n_0 \approx e^{-12} \approx 10^{-5}$

$$(k_B = 1.380650 \times 10^{-23} \text{ J/K and 1 eV} = 1.602 \times 10^{-19} \text{ J})$$

This would mean that there would be 10 vacancies for every million sites.

Schottky Defect

In ionic crystals, it is usually found that roughly the same number of positive and negative ion vacancies is formed. The formation of a pair keeps the crystal electrostatically neutral. This pair is called the Schottky defect.

From statistical calculation, we can obtain

$$n \approx N \exp^{(-E_p/2kT)} \tag{4.4}$$

for the number of pairs where E_p is the energy of formation of pair and N is the number of atoms.

Frenkel Defect

A small displacement of an atom from a normal site to a nearby normally unoccupied site will simultaneously introduce both a vacancy and an interstitial defect (Figure 4.1). Such a vacancy interstitial pair is called a Frenkel defect. Thus, a Frenkel defect ensures electrostatic neutrality. The presence of extra strain at the locations leads to an increase in the strain energy in the crystal and a disorder around them. However, the two positions need not be necessarily adjacent. If the number of Frenkel defects n is smaller than the number of lattice site N and the number of interstitial sites N', the result is

$$n \approx (NN')^{1/2} e^{(-E_t/2kT)} \tag{4.5}$$

where E_t is the energy necessary to move the atom from lattice site to an interstitial position.

Interstitial Defect

The converse of absence of an atom from a normally occupied site would be the presence of an atom in a normally unoccupied site, that is, a site in between normal lattice sites. Such an atom is called an interstitial. While we might expect a vacancy to cause a local increase of atomic separation, with the surroundings slightly in tension, the introduction of an interstitial will usually cause local squeezing and compression of the lattice. An interstitial may be self-interstitial wherein the same atom belonging to the lattice is misplaced in the interstitial position or it may be a foreign atom in the interstitial position. Interstitial impurity atoms are usually much smaller in size than the atoms of the bulk matrix. Both of these defects can be created by thermodynamic fluctuations or other effects resulting in the dislodging of an atom from its original site, which goes to either the surface or an interstitial site.

Substitutional Defect

Another type of point defect is an impurity atom, an atom of an abnormal type at a normal lattice site. Such an atom is called a substitutional defect. The impurity might be an atom of a different chemical element or a different isotope of the normal element or even an atom of the normal element in a different state of ionization or magnetization. Any of these circumstances is a deviation from the ideal and hence results in a defect. The substitutional atoms are usually close in size (within ~15%) to the atom in the bulk.

All these defects are schematically shown in Figure 4.1. Study of ionic conductivity is helpful in understanding the nature of the aforementioned lattice defects, particularly in ionic crystals.

Color Centers

Finally, we may note that more complicated point defects may be built up by the combinations of some of these simpler ones. The most famous example of this is the color center, which is a defect responsible for the coloration in many crystals. Pure alkali halide crystals are transparent throughout the visible region of the spectrum. But they can be colored in a number of ways such as (1) introduction of chemical impurities such as transition elements, (2) introduction of excess neutral Na in NaCl crystal (crystal turns yellow), (3) introduction of excess neutral K in KCl (crystal turns magenta), and (4) irradiation with x-rays, γ rays, neutrons, or electrons.

The most common of the color centers is the F center. The name *F center* comes from the German word *Farbe*, which means "color." The F center has been identified by electron paramagnetic resonance (EPR) as an electron bound to a negative ion vacancy. If NaCl is heated in an atmosphere of sodium vapor, it becomes slightly nonstoichiometric with excess sodium atoms in the lattice. The additional sodium atoms require the formation of vacancies in the chlorine lattice, and the extra electrons absorbed by the sodium atoms are soon captured by the chloride ion vacancies, thereby forming an F center.

An antimorph of F centers is a hole trapped at positive ion vacancy, but no such center has been identified in alkali halides. The best known trapped hole center is V_k center. V_k center is found where a hole is trapped by a halogen ion in an alkali halide crystal. It is also possible to produce small clusters of color centers (two or three together), which have different optical properties.

It should be remembered that point defects are not static but move around the lattice randomly through thermal diffusion. Diffusion in solids, however, is pretty slow at room temperature. But it increases exponentially with temperature.

While these point defects could explain physical phenomena like diffusion process in solids, electrical conductivity, luminescence, and so on, they cannot account for phenomena associated with plastic deformation.

Line Defects or Dislocations

Historically, the presence of dislocations in crystal was inferred from the disparity between the theoretically calculated and experimentally observed critical shear stresses required to cause plastic deformation. In this section, we see how this disparity could be reconciled by the introduction of dislocations in the crystalline lattice.

When a crystal is deformed elastically under the influence of applied stresses, it returns to its original state upon removal of the stresses. However, if the applied stresses are sufficiently large, a certain amount of deformation remains even after the removal of stresses, and the crystal is said to be plastically deformed. This property (i.e., the plasticity of pure single crystals of many solids) is a striking feature of their mechanical behavior. In many crystals, the plastic flow results from the sliding of one part of the crystal relative to the other. The sliding process is referred to as slip. The plane and direction in which the slip occurs define, respectively, the slip plane and slip direction. The visible intersection of the slip planes with the outer surface of the crystal is known as slip band (Figure 4.2) [1]. The amount of slip associated with a band may be several thousands of lattice units.

One of the important aspects of slip is its inhomogeneous nature in the sense that only a relatively small number of atoms actually take part in the process; that is, only those atoms which form layers on either side of the slip plane (Figure 4.2). Elastic deformation, on the contrary, displaces all the atoms in the crystal. This difference between the plastic and elastic deformation indicates that the atomic interpretation of plastic flow must be based on an entirely different model than that of the elastic deformation.

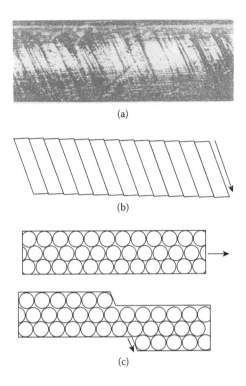

FIGURE 4.2
(a) Macroscopic slip bands. (From Rosenberg, H. M., *The solid state, an introduction to physics of crystals*, 2nd ed, 1978, Clarendon Press, Oxford, by permission of Oxford University Press.); (b) schematic of figure in (a); and (c) schematic representation of the inhomogeneous nature of the slip.

In fact, elastic properties of solids can be understood quite well in terms of interatomic forces acting in a perfect lattice. Plastic deformation, however, cannot be discussed properly on the basis of a perfect lattice.

Besides being characterized by inhomogeneity, plastic flow is also anisotropic. Slip usually takes place preferentially in planes of high atomic density. For example, slip occurs along (111) planes in a face-centered cubic (fcc) lattice. Also, the direction of slip commonly coincides with the direction along which the number of atoms per unit length is high. Another important result obtained from experiments is the existence of critical shear stress. In other words, the slip obeys Schmid's law of critical shear stress, which states that the slip takes place along a given slip plane and direction when the corresponding component of shear stress reaches a critical value.

In Figure 4.3, a force F is applied along the length of the crystalline rod having a cross-sectional area of A. The normal to the slip plane makes an angle of φ with the direction along which the force is applied, and slip direction itself makes an angle of θ with it. The component of force in the direction

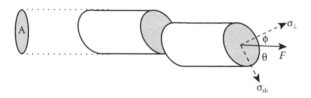

FIGURE 4.3
Resolution of applied stress into σ_{sh}, the component parallel to the slip plane in the direction of slip.

of the slip, therefore, is $F \cos\theta$. The area of the slip plane is $A/\cos\varphi$. Therefore, shear stress σ_{sh} is given by

$$\sigma_{sh} = (F/A) \cos\theta \cos\varphi \tag{4.6}$$

Theoretical Estimation of Critical Shear Stress

There is a simple way, according to Frenkel [2], to estimate the maximum shear stress that the lattice can withstand. We choose a lattice plane as a glide/slip plane and in it a row of atoms as the direction of gliding. Let d be the distance between the two neighboring lattice planes and a the distance between two atoms in the slip direction (Figure 4.4a). If a shear stress σ acts in the slip plane parallel to the slip direction, two neighboring planes suffer a relative displacement x. The shear strain is thus x/d. For small values of x/d, Hooke's law holds, and we have

$$\sigma = G(x/d) \tag{4.7}$$

where G is the shear modulus.

In Figure 4.4b, where σ is plotted as a function of x, Equation 4.7 is represented by the dashed line passing through the origin.

If we assume that the distortion of the lattice remains strictly homogeneous, however large the strain is, we see that for $x = a/2$ the stress σ must vanish since all the atoms of a plane are symmetrically situated with respect to atoms of the neighboring planes and hence must be under equilibrium. For $x = a$, we have again an undistorted lattice and the relationship between σ and x in the neighborhood of $x = a$ is given by a straight line representing Hooke's law. Further, on both sides of the equilibrium sites, the forces differ only in sign. Thus, σ may be approximated to a sine function of x; that is,

$$\sigma \approx \sigma_m \sin\left(\frac{2\pi x}{a}\right) \tag{4.8}$$

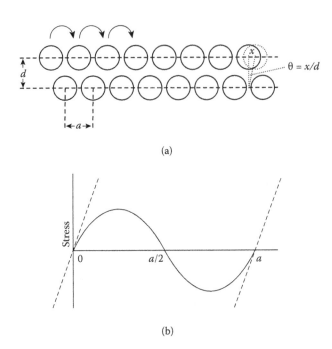

(a)

(b)

FIGURE 4.4
(a) Calculation of shear stress to cause simultaneous slip in an ideal lattice and (b) shear stress
as a function of relative displacement approximated to sinusoidal function.

This is shown in Figure 4.4b as a full line. Evidently, σ_m is the maximum
shear stress that the lattice can withstand, and this in turn is the theoretical
shear strength of the glide system considered.

For small values of x/a, Equation 4.8 becomes

$$\sigma \approx \sigma_m \left(\frac{2\pi x}{a} \right) \tag{4.9}$$

This equation must be same as Equation 4.7 and hence comparing the constants, we get

$$\sigma_m \approx \frac{G}{2\pi} \tag{4.10}$$

where we have assumed $a \approx d$.

This means the theoretical shear strength is approximately one-sixth of
the shear modulus. However, this quantity for single crystals of common
metals is several orders of magnitude greater than experimentally obtained
macroscopic yield stresses.

Refinement in Calculation of Critical Shear Stress

The adoption of the sinusoidal relation in the earlier calculation implies that the critical shear stress is reached when the atoms are displaced by a distance $a/4$ in the slip direction. It was pointed out that the critical shear stress is overestimated by this assumption and must be less than this when a realistic law is taken for the forces between atoms [3].

Another point to be noted is that the calculation of theoretical stress takes no account of possible configuration of mechanical stability through which the lattice may pass as it is sheared. The fcc metals, for example, pass through twinning and then body-centered cubic configurations when sheared on their slip planes. In such cases, the force-displacement relation will oscillate over smaller periods than the lattice spacing, and the critical shear stress will be further reduced.

Frenkel's calculation was refined by Mackenzie [4] by taking account of both of the effects. Using reasonable laws of interatomic forces and allowing for all intermediate positions of stability in the sheared lattice, he showed that the theoretical shear stress can be reduced to about $G/30$ but not any further. This is still immensely large compared to the observed strength of soft crystals. Since G is of the order of 10^{10} N/m^2, one obtains a value of 10^8–10^9 N/m^2 for σ_c on the basis of the earlier model. However, in practice, for pure crystals the critical shear stress is in the range of 10^5–10^6 N/m^2, which is several orders of magnitude less than the theoretical one. For example, in the case of aluminum single crystals, the measured shear modulus is 2.5×10^{10} N/m^2, where the elastic limit is only 4×10^5 N/m^2, giving a value of ~60,000 for G/σ_m [5]. The deviation cannot be explained by the thermal motion because measurement carried out at very low temperatures also yields surprisingly low critical shear stress. Hence, it is clear that the shear stress cannot be treated simply by extending the theory of elasticity to large strains. Since the crystalline nature of the solid bodies is a proven fact, the error can only lie in the assumption that crystals are perfect and the supposition that the total glide takes place at the same time along the entire slip plane.

In contrast to what has been assumed to obtain an expression for the critical shear stress, the process of plastic deformation is presumed to consist of slip that does not take place simultaneously in the whole slip plane. Hence, in the slip plane, there will be slipped and unslipped areas. The boundary between these two areas clearly creates another type of crystal defect. The boundary is either a closed loop or a half loop extending from surface to surface that sweeps along the glide plane during deformation. This new type of defect is essentially a line defect and is called dislocation. It is also obvious from this definition that a dislocation line cannot end in the middle of a crystal.

Two simple types of dislocations are realizable, with these being called edge and screw dislocations. However, in general, a dislocation may be a mixture of these two types and hence called mixed dislocation. The geometry of mixed dislocation is rather difficult to describe, but the structure of pure edge and screw types can be easily illustrated.

Edge Dislocation

The arrangement of atoms around an edge dislocation is sketched in Figure 4.5a for a simple cubic structure. Here, the distortion of the crystal structure may be due to the insertion of an extra half plane of atoms into the lower part of the crystal. Most of the distortion is concentrated at the upper edge of this half plane of atoms all along its length—hence the name "edge dislocation." Figure 4.5b is the sketch of edge dislocation in three dimensions. It is clear from this figure that dislocation is a line imperfection in contrast to a point defect that we discussed earlier. In the case of an edge dislocation the atoms below the edge of the extra half plane are under compression and those above are under tension. It may be noted that insertion of integral number of half planes would result in a dislocation of multiple strength.

One can also visualize the formation of an edge dislocation in an alternate way. Imagine a perfect cubic lattice to begin with (Figure 4.6a). Let a stress be applied to the upper part of the crystal along the slip direction so that this part of the crystal begins to slip with respect to the lower part along the slip plane. This process results in a boundary between the slipped and unslipped region, which by our definition is a dislocation (Figure 4.6b). The atomic arrangement around the dislocation suggests that it is of edge type. As we continue to apply the stress, we may suppose that this boundary moves to the right until eventually the whole of the upper part has slipped over the lower part as shown (Figure 4.6c). The dislocation itself has now vanished, and it has passed through the crystal. Thus, we can imagine that the stress applied has been used initially to create a dislocation and then to move it across the entire crystal. We can also repeat the entire process

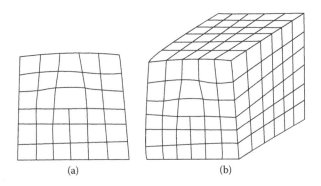

(a)   (b)

FIGURE 4.5
Edge dislocation: (a) two- and (b) three-dimensional views.

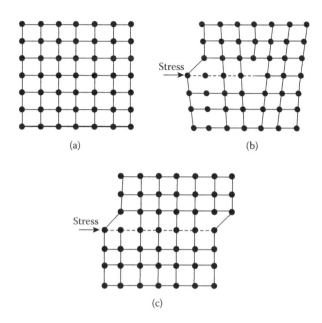

(a)

(b)

Stress

(c)

FIGURE 4.6
Creation and motion of an edge dislocation during the slip. (a) perfect two dimensional lattice;
(b) applied stress has created an edge dislocation in the lattice and moved it to the middle and
(c) on further application of the stress, the dislocation has moved out of the lattice and the slip
is complete.

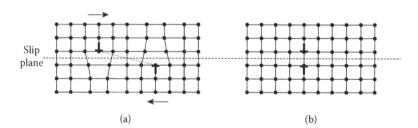

(a)

(b)

FIGURE 4.7
Annihilation of (a) positive and (b) negative dislocations.

by applying the stress to the lower portion of the crystal in the opposite
direction so that the lower part of the crystal moves from right to left. Then,
we would have created a dislocation in the lower part of the crystal that
eventually moves from right to left. These two configurations are conven-
tionally designated as positive and negative dislocations. As always with
the convention, we may assign the sign of the dislocation as we like as
long as we are consistent. While dislocations of opposite signs attract and
annihilate when the magnitude of slip vector is the same (Figure 4.7), those
of the same sign repel each other. If we assume in Figure 4.7 the boundary

created to be straight and perpendicular to the plane of the paper, we see that the direction of slip is perpendicular to the dislocation line.

Screw Dislocation

The other type of dislocation called screw dislocation is sketched in Figure 4.8. The features of a screw dislocation can be understood by the following operation on a perfect crystal. Let a sharp cut be made halfway through a perfect crystal. Let the material on one side of the cut be sheared up relative to the other by an atomic spacing as shown in Figure 4.8. Clearly a line of distortion exists along the edge of the cut. This line is called a screw dislocation. Again, shearing can be integral multiples of the atomic spacing.

The most important aspect of the screw dislocation is the new feature of the atomic planes around this dislocation. In this configuration, complete planes of atoms normal to the dislocation line do not exist. Rather, all atoms lie on a single surface, which spirals from one end of the crystal to the other end and hence the word *screw*. Unfortunately, it is a little more difficult to visualize the atomic arrangement around this dislocation. Nevertheless, this line is the line of demarcation between the slipped and the unslipped regions, the slip occurring this time parallel to the dislocation line. It may be noted that even in the case of screw dislocations, we can designate positive and negative character to them depending on the direction of slip vector. The positive and negative nature of dislocations has relevance to crystal growth.

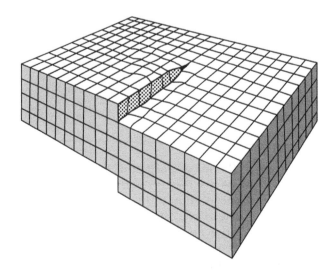

FIGURE 4.8
Schematic of a screw dislocation.

Mixed Dislocation

In the two extreme cases just discussed, we see that the dislocation lines are straight and the slip direction is either parallel (for screw type) or perpendicular (for edge type) to the dislocation line. However, in general, dislocations need not be straight and the slip vector may be inclined to the dislocation line. Such dislocations are called mixed dislocations. In fact, the same dislocation line can exhibit screw, edge, and mixed character, as shown in Figure 4.9. As can be seen, this dislocation exhibits edge character on face A and screw character on face B. Along the line wherever the slip direction is inclined to it (e.g., at point C), it exhibits mixed character with both screw and edge components. But what has remained constant throughout is the magnitude and direction of slip vector.

Burgers Vector

It is clear from the earlier discussion that the direction and magnitude of the slip vector is the true character of a dislocation. In fact, it is the fundamental quantity that defines an arbitrary dislocation. This vector is known as the Burgers vector **b**. Its magnitude (sometimes called the strength of dislocation) is integral multiple of the repeat vector of the lattice and is usually the smallest one so that in simple crystals it is of the order of interatomic spacing. Also, the Burgers vector lies in the slip plane and is perpendicular to an edge dislocation, parallel to a screw dislocation, and inclined to a dislocation of mixed character.

Burgers vector's atomistic definition follows from a Burgers circuit around the dislocation in the real crystal, which is illustrated in Figure 4.10. If we go along a closed circuit from lattice point to lattice point in an ideal lattice, we obtain a closed chain of the base vectors that define the lattice.

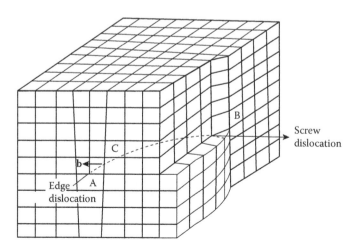

FIGURE 4.9
Geometrical representation of a mixed dislocation.

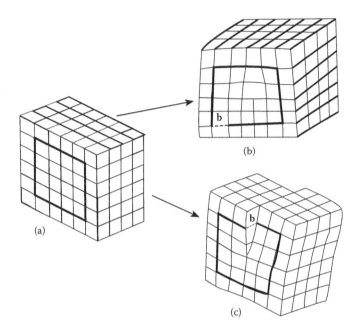

FIGURE 4.10
Burgers circuit around (a) a perfect lattice, (b) an edge dislocation, and (c) a screw dislocation.

However, if we go along exactly the same way enclosing a dislocation as shown, it will not close. The special vector needed for closing the circuit is by definition the Burgers vector **b**. It follows that the Burgers vector of a perfect dislocation is necessarily an integral multiple of lattice vector. We also have to define if we go clockwise or counter clockwise along the Burgers circuit. In both cases we will always get the same Burgers vector, but the sign will be different. Although the atomistic representation of screw dislocation is somewhat difficult to draw, a Burgers circuit can still be drawn around it, as shown in Figure 4.10.

Motion of Dislocations

We have seen earlier that dislocations can move in the lattice. Predictably, dislocation motion is more constrained as compared to point defects. Two types of motion are possible in general. They are called slip and climb.

The type of motion shown earlier for an edge dislocation occurs in the slip plane. Here, the dislocation can move rather easily either to the right or to the left along the slip plane. Such a motion is called conservative motion because when the dislocation moves, there is neither an addition nor a deletion of atoms from the half plane. There is only a shuffling motion of the atoms in the core of the dislocation as it moves. This type of motion is referred to as slip or glide. Glide motion is also possible for a screw dislocation. However, since all

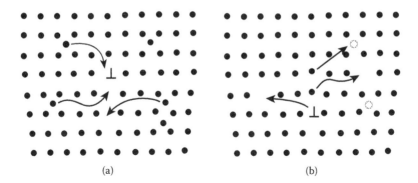

FIGURE 4.11
Climb of an edge dislocation by the (a) addition and (b) subtraction of atoms at the half plane.

the planes containing screw dislocation also contain Burgers vector, screw dislocation can move in any direction by glide. Also, since the ledge (step) formed by the presence of screw dislocation appears on the surfaces of the crystal where the dislocation ends, the ledge has to follow the dislocation as it moves.

An edge dislocation can also be moved vertically upward or downward from the slip plane by the deletion or addition of atoms, respectively, at the edge of the half plane. Consequently, both these types of motion are nonconservative and the motion as such is referred to as climb. In an actual crystal, dislocation does not climb by moving the extra plane bodily. The same effect can, however, be realized if extra atoms are added to or removed from the half plane individually. This is possible by the diffusion of interstitial atoms (climb down) or vacancies (climb up) toward the edge of the half plane (Figure 4.11). The dislocation climb is more probable at high temperatures because extra energy is required to produce and diffuse these point defects toward or away from dislocation. Since pure screw dislocations do not have extra half planes, they do not climb the way edge dislocations do. They can move conservatively in any slip plane.

Energy of a Straight Screw Dislocation

An important characteristic of all dislocations is that the lattice distortion is severe only in the immediate vicinity of the dislocation line. In fact, along the dislocation line the atoms may not even have the correct coordination number corresponding to the lattice in which they are formed. However, a few lattice distances away from the dislocation line, the distortion is so small that the lattice is nearly perfect and can be considered as an elastic continuum. The region near the dislocation line where the distortion is extremely large, and hence the local strain is extremely large, is called the core of the dislocation. Estimation of the energy of this core is quite involved and has been carried out only for some simple structures. However, it has been found that, in general, the contribution by the core to the total energy of dislocations would not exceed 10%. Hence, one normally estimates only the elastic energy of the dislocation.

The elastic energy of dislocation can be estimated by assuming that the crystal behaves like an elastic solid during the process of creation of dislocation. Let us consider the cut made in a cylindrical crystal of length l and radius r, as shown in Figure 4.12a. The creation of a straight screw dislocation at the axis of this cylinder amounts to a displacement parallel to the cut. To cause this displacement, forces must be applied over the surface of the cut, and the work done by the forces in making the displacement b would be equal to the energy E_D of the produced dislocation.

Hence,

$$E_D = \int F \cdot b \, dA \tag{4.11}$$

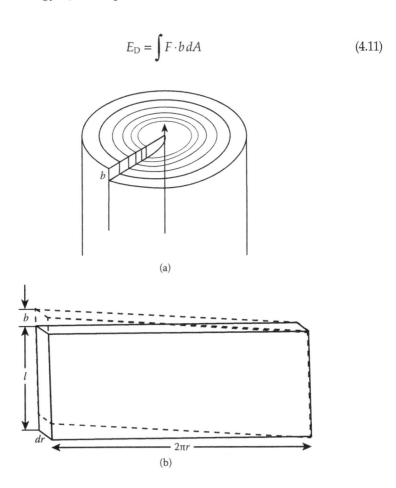

(a)

(b)

FIGURE 4.12
(a) Creation of a screw dislocation at the axis of a cylindrical crystal and (b) the outer most cylindrical shell around the screw dislocation opened out into a sheet.

where the integral is evaluated over the entire area of the surface of cut. The force F is the average force per unit area at a point on the surface during the displacement. The average value is used because the force at a point builds up from zero to a maximum value as the displacement is carried out, and we presume that it builds up linearly.

To calculate the energy, we should evaluate the magnitude of F. For this purpose, let us assume that this crystal is made up of concentric cylindrical shells with the dislocation running at their center (Figure 4.12a). If the outermost cylindrical shell is opened out, it would appear like a platelet of length l, width $2\pi r$, and thickness dr, as shown in Figure 4.12b. Now the shell is sheared with a force $f \, l \, dr$ along the edge of the plate to produce a displacement b, where f is the stress. Since for small displacements Hooke's law is obeyed, the stress f to cause a shear b in a cylindrical shell at a distance r from the center can be written as

$$f = \frac{Gb}{2\pi r} \tag{4.12}$$

where $(b/2\pi r)$ is the shear strain and G is the shear modulus of the material.

Now the average force per unit area can be taken as just half the final value when the displacement is b. Hence

$$F = \langle f \rangle = \frac{1}{2} f = Gb/4\pi r \tag{4.13}$$

Substituting this in Equation 4.11, we get

$$E_D = \int \frac{Gb}{4\pi r} \cdot b \, dA = \frac{Gb^2}{4\pi} \int \frac{dl \, dr}{r} = \frac{Gb^2}{4} \int_{r_0}^{r} \frac{dr}{r} \int_{0}^{l} dl$$

$$= \frac{Gb^2 l}{4\pi} \ln \frac{r}{r_0} \tag{4.14}$$

Hence, dislocation energy per unit length is

$$\frac{Gb^2}{4\pi} \ln \frac{r}{r_0} \tag{4.15}$$

where r and r_0 are appropriate upper and lower limits of the variable r. Obviously, the upper limit of the variable cannot exceed the dimensions of the crystal and more often much smaller because crystals contain many dislocations and these dislocations in random distribution cancel out each other's stain fields at distances approximately equal to the mean distance between them. The reason for choosing a lower limit r_0 is not obvious. As you can see from Equation 4.15, as $r_0 \to 0$, expression becomes divergent and one cannot evaluate the energy of dislocation unless it is made to converge. Also, it is important to realize that the extension of elastic theory to the core region of dislocation where the atomic displacements are sufficiently large would not be correct. The reasonable lower limit therefore corresponds to the distance from the dislocation line from where the elastic law can be applied. The value of r_0 would be of the order of lattice spacing from the geometric center of the dislocation or Burgers vector.

The calculation for the energy of edge dislocation is more complicated because both shear and normal stresses are present. Since in the case of an edge dislocation the region around the extra half plane is under compression and that below it is under tension, the calculation of energy would also involve Poisson ratio of the crystal. Detailed calculations are available [6,7], and the resulting formula for an isotropic crystal is as follows:

$$E = \frac{Gb^2}{4\pi(1-v)} \ln \frac{r}{r_0} \qquad (4.16)$$

where v is the Poisson ratio ($v \sim 0.3$ for most crystals) and other symbols have their usual meaning. Thus, the strain energy of edge dislocation is slightly higher than that of screw dislocation for the same Burgers vector.

From these expressions, we can see that the energy of a dislocation is proportional to the square of its Burgers vector, and a first approximation of energy per unit length can be taken to be

$$E_D \sim Gb^2 \qquad (4.17)$$

This tells us why dislocations of multiple strength are not energetically favorable, because the system will attain lower configuration energy by splitting into several dislocations, with each having as small a b value as possible. Tendency to minimize the value of b also implies that there will be repulsion between like dislocations on the same slip plane.

One of the important characteristics of dislocations is the fact that they are nonequilibrium defects. Unlike vacancies, which at thermal equilibrium exist in significant numbers given by the Boltzmann factor, dislocations in crystals seem to have zero equilibrium density. This means, in principle, one can grow crystals free of dislocations. The experimentally observed dislocation densities, however, depend on many factors, such as method of growth, previous thermal history, previous mechanical treatment, and so on.

The Force on a Dislocation When a Shear Stress Is Applied

We saw earlier that the stress applied on a perfect crystal to cause slip resulted in the creation of a dislocation and its motion through the crystal. Now, if dislocations are already present in the crystal, then the process of slip would involve only their motion. We should now demonstrate that the force required to move a dislocation is very small as compared to that required to form it so that we can reconcile with the discrepancy between the theoretically estimated and experimentally determined values for the critical shear stresses. Unfortunately, a precise calculation of this is quite difficult because these energies are very sensitive to the atomic arrangement. However, a rough estimation of these can be done, which is given in this section.

Let the stress σ move the dislocation from one end to the other, that is, a distance x. Referring to Figure 4.13, when this occurred, the top portion of the specimen would have moved a distance b relative to the bottom; hence, the work done by σ is

$$\text{Work done} = \text{force} \times \text{displacement} = \sigma\,x\,y\,b \qquad (4.18)$$

where σ is the stress, $x\,y$ is the area, and b is the displacement.

This must in principle be equal to the work done on the dislocation when it is moved by a distance x.

Now, if F is the force per unit length of dislocation, this work is $F\,x\,y$ so that

$$F\,x\,y = \sigma\,x\,y\,b \qquad (4.19)$$

Thus, the force on dislocation is

$$F = \sigma\,b \qquad (4.20)$$

Now, if a unit length of dislocation is moved through a distance b, the work done, ΔE, is given by

$$\Delta E = \sigma\,b\,b = \sigma\,b^2 \qquad (4.21)$$

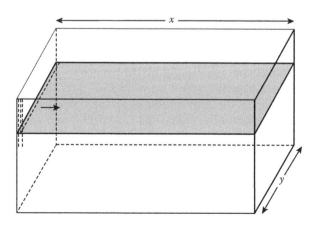

FIGURE 4.13
Motion of a straight dislocation in the slip plane.

This energy is a measure of the extra energy required to move a dislocation. But the elastic energy of dislocation E_D is $\dfrac{Gb^2}{4\pi}\ln\dfrac{r}{r_0}$ and if r_0 is taken as 10^{-10} m and $r = 10^{-2}$ m, the elastic energy would be of the order of Gb^2 per unit length. Hence

$$\frac{\Delta E}{E} \approx \frac{\sigma b^2}{Gb^2} = \frac{\sigma}{G} \tag{4.22}$$

Experimental evaluation of this quantity yields $\approx 10^{-5}$. Hence, the extra energy involved in getting the dislocation moved is only 10^{-5} of the elastic energy of dislocation.

Now we can understand why experimental values of critical shear stress measured for various crystals are so low compared to the theoretical estimates. These crystals were all with dislocations, and the stress applied was essentially moving dislocations to cause slip. Had the crystals been dislocation-free, the experimental values of critical shear stresses would have approached the theoretical values.

Dislocation Multiplication

We mentioned earlier that the process of slip involves movement of dislocation in the slip plane. For a macroscopic slip band to occur, a large number of dislocations have to sweep across the crystal in succession. This would happen through dislocation multiplication followed by their motion. How do dislocations multiply? One possible mechanism by which dislocations can multiply has been suggested by Frank and Read [8], and this mechanism is illustrated schematically in Figure 4.14.

Let us consider a segment of dislocation line that is firmly anchored at two points A and B. Such an anchoring is possible at impurity sites, at nodes in a dislocation network, or at some other crystal defects. Under the application of a suitable stress, the anchored dislocation tends to bow out. As the dislocation sweeps the slip plane, the plane above undergoes a relative displacement of b with respect to the slip plane. Eventually, the dislocation will become semicircular, at which point it would start developing lobes around the points A and B. As the application of stress continues, the two lobes would expand and eventually the points C and D touch each other. Since points C and D belong to the same dislocation but move in opposite direction, they would annihilate each other to form a perfect lattice in the region of contact. In doing so, they leave behind a closed dislocation loop and the original anchored section of the dislocation. This process can repeat over and over again so that a whole series of dislocation loops are formed one inside the next. The process will, however, stop when the outermost dislocation loop encounters a barrier that prevents further expansion. Since like dislocations repel each other in the same plane, the barrier exerts a backward

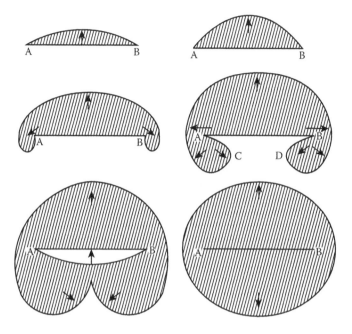

FIGURE 4.14
Dislocation multiplication by Frank-Read mechanism.

stress on all the inner loops, which in turn opposes the applied stress, and the Frank-Read source finally ceases to operate.

Low-Angle Grain Boundaries

One should realize that even good single crystals can have small misorientations of the lattice between adjacent parts of the crystal, and such crystals are said to be mosaic. The boundary between these two slightly misoriented parts is known as a low-angle grain boundary. It was suggested by Burgers [9,10] that the low-angle boundaries in crystals consist of an array of dislocations. This is illustrated in Figure 4.15, in which the boundary occupies a (010) plane of a simple cubic lattice and divides the two parts of the crystal having [001] axis in common. This is referred to as a pure tilt boundary because the existing misorientation can be described by a small rotation θ with respect to the common axis [001] of one part of the crystal relative to the other. As you can see from Figure 4.15, the gaps between the two parts of the crystal are filled by extra half planes at equal intervals. We immediately notice that these extra half planes form edge dislocations where they terminate; hence, a low-angle grain boundary may be considered to be made up of an array of edge dislocations. Dislocations thus formed are of the same sign and have a spacing $D = b/\theta$ between them, where b is the magnitude of the Burgers vector. Because they are of same sign, dislocations stacked one

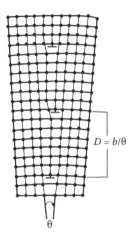

FIGURE 4.15
Schematic of a low-angle grain boundary.

above the other in this fashion lead to a stable configuration. Now, if θ is only 2 arc seconds (~10^{-5} radian), for a Burgers vector of 5Å (0.5 nm), the average spacing will be ~5×10^{-5} m.

On the contrary, if two portions of a crystal are rotated through a small angle about an axis perpendicular to the resulting low-angle grain boundary, a simple twist boundary is produced. This boundary would consist of array of screw dislocations forming a grid. As such low angle grain boundaries can also be categorized as planar defects.

Planar Defects

For the sake of completeness, we briefly introduce these defects. However, detailed discussion on them is beyond the scope of this book, and the readers are advised to refer to an excellent book written by Verma and Krishna [11].

A disruption of the long-range stacking sequence can produce two other common types of crystal defects: a stacking fault and a twin region. A change in the stacking sequence over a few atomic spacing produces a stacking fault. However, a change occurring over many atomic spacing produces a twin region. Both of these defects are essentially planar in nature.

Stacking Faults

A stacking fault is a one- or two-layer interruption in the stacking sequence of atomic planes. Stacking faults occur in a number of crystal structures, but it is easy to visualize how they occur in close-packed structures.

For example, it is known that fcc structures differ from hexagonal close-packed (hcp) structures only in their stacking order. For hcp and fcc structures, the first two layers arrange themselves identically, to give an AB arrangement (Figure 4.16). If the third layer is placed so that its atoms are directly above those of the first layer, the stacking will be ABA. This is the hcp structure, and the layer sequence continues as ABABABAB. However, it is possible for the third layer of atoms to arrange themselves so that they are neither directly above the first layer nor above the second layer but at a third equivalent position to produce an ABC arrangement. This is the fcc structure. These three positions are shown in Figure 4.16. Now, if the layer sequence in hcp structure is suddenly switched to that of an fcc sequence (e.g., ABABABCABAB), a stacking fault is produced. Alternately, in the fcc arrangement where the stacking pattern is ABCABCABC, a stacking fault would appear if one of the C planes is missing. In other words, the sequence would become ABCABCAB_ABCABC.

Above stacking faults can be produced either by the agglomeration of vacancies or by the agglomeration of self-interstitials in the stacking plane. For historical reasons stacking fault produced by vacancy agglomeration is called intrinsic stacking fault and that produced by interstitial agglomeration is called extrinsic stacking fault. A disc of vacancies (or self-interstitials) produced to create a stacking fault obviously is bordered by an edge dislocation. Now, if we construct a Burgers circuit around such a dislocation, we will realize that the closer failure (i.e., the Burgers vector) is not a translation vector of the lattice. For example, for an fcc lattice, it would be $(a/3)<111>$. Dislocations with Burgers vector of this type are called partial dislocations. For a detailed understanding of stacking fault and their interconnection with partial dislocations, the reader is advised to refer to the book mentioned earlier [11].

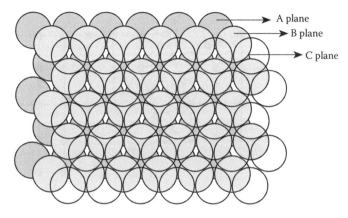

FIGURE 4.16
Stacking of A, B, and C planes in a close-packed structure.

Twin Boundaries

The crystals are often produced with faults in which one region of the crystal is a mirror image of the other. The atoms in one part of the crystal are in positions that can be produced by reflecting the atoms in the second part at some symmetry plane of the crystal (Figure 4.17). This is known as twinning.

Twinning often occurs in metals that have small stacking fault energy. This implies that the extra energy required for a small atomic mismatch is small. If a stacking fault does not correct itself immediately but continues over some number of atomic spacing, it will produce a second stacking fault that is the twin of the first one.

Twinned crystals can be formed in three ways: (1) It can be formed during growth (growth twins) as a result of an interruption or change in the lattice during formation, or growth due to a deformation caused by a larger substituting ion; (2) Annealing or transformation twins are due to a change in the crystal system during cooling as one form becomes unstable and the crystal structure reorganizes (phase transformation) into another more stable form; and (3) Deformation or gliding twins are the result of stress on the crystal after the crystal has been formed. During crystal growth, crystals that grow adjacent to each other may be aligned to resemble twinning. This parallel growth simply reduces system energy and should not be confused with twinning.

Twinning can often be observed by optical microscopy under crossed polarizers. The presence of twins can also be detected by x-ray diffraction since an extra set of diffraction spots are produced from the twinned regions.

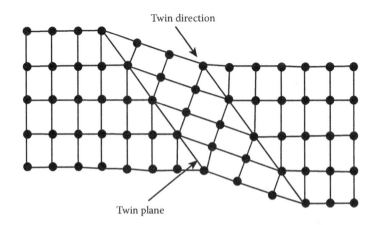

FIGURE 4.17
Twin plane and twin direction.

FIGURE 4.18
Grains and grain boundaries seen in an axial cross section of cadmium telluride crystal boule.
Marker represents 5 mm. (From Thomas, R. N., et al., *Journal of Crystal growth* 99, 643–653, 1990.
With permission.)

Grain Boundaries in Polycrystals

Up to this point, we were dealing with defects in single crystals. However,
solids in general and metals in particular are polycrystalline, consisting of
a number of crystallites or grains (Figure 4.18) [12]. Grains can range in size
from nanometers to millimeters across, and their orientations are usually
rotated with respect to neighboring grains. One grain meets another grain
at a grain boundary, and these are planar defects. Since grain boundaries
limit the lengths and motions of dislocations, a material having smaller
grains (more grain boundary surface area) is usually stronger compared
to the same material with larger grains. The size of the grains can be
controlled by the cooling rate when the material is cast or heat treated.
Generally, rapid cooling produces smaller grains, whereas slow cooling
results in larger grains.

Bulk Defects

Bulk defects are three dimensional and occur in crystals on a much bigger
scale than the rest of the defects discussed so far. One of the most common
volume defects encountered is the void. Voids are regions within the crystal where a large number of atoms are missing from the lattice. These can
occur for a variety of reasons. For example, voids occur when air bubbles get

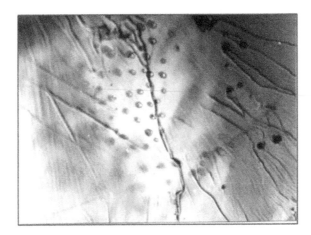

FIGURE 4.19
Volume defect: inclusions (100×). (From Bhat, H. L., Studies on growth and defect properties of barite group crystals, PhD dissertation, SPU University, Vallabh Vidyanagar, India, 1973.)

trapped during the solidification process. They can also occur due to shrinkage of the material when it solidifies (cavitation).

Another type of volume defect is the inclusion. This occurs when impurity atoms cluster together to form a small region of different phase. These are also called precipitates. Also, during crystal growth from solution, the mother liquor can get trapped to form inclusions. This is shown in Figure 4.19 for a naturally grown $SrSO_4$ crystal [13].

References

1. Rosenberg, H. M. 1978. *The solid state, an introduction to physics of crystals*, 2nd ed. Oxford, UK: Clarendon Press.
2. Frenkel, J. 1926. Zur Theorie der Elastizitätsgrenze und der Festigkeit kristallinischer Körper. *Zeitschrift für Physik* 37: 572–609.
3. Cottrell, A. H. 1953. *Dislocations and plastic flow in crystals.* Oxford, UK: Clarendon Press.
4. Mackenzie, J. K. 1949. A theory of sintering and the theoretical yield strength of solids. PhD dissertation, University of Bristol, Bristol.
5. Kittel, C. 1971. *Introduction to solid state physics*, 4th ed. New York, NY: John Wiley, p. 672.
6. Friedel, J. 1964. *Dislocations*. New York, NY: Pergamon Press.
7. Read, W. T. 1953. *Dislocations in crystals*. New York, NY: McGraw-Hill.
8. Frank, F. C., and W. T. Read. 1950. Multiplication processes for slow moving dislocations. *Physics Review* 79: 722–723.

9. Burgers, M. J. 1939. Some considerations on the fields of stress connected with dislocations in a regular crystal lattice. *Koninklijke Nederlandse Akademie van Wetenschappen* 42: 293–325.

10. Burgers, M. J. 1940. Geometrical considerations concerning the structural irregularities to be assumed in a crystal. *Proceedings of the physical Society* 52: 23–33.

11. Verma, A. R., and P. Krishna. 1966. *Polymorphism and polytypism in crystals.* New York, NY: John Wiley.

12. Thomas, R. N., H. M. Hobgood, P. S. Ravishankar, and T. T. Braggins. 1990. Meeting device needs through melt growth of large-diameter elemental and compound semiconductors. *Journal of Crystal Growth* 99: 643–653.

13. Bhat, H. L. 1973. Studies on growth and defect properties of barite group crystals. PhD dissertation, SPU University, Vallabh Vidyanagar, India.

5

Dislocations and Crystal Growth

Screw Dislocation Theory of Crystal Growth

In Chapter 3, we saw that growth under very low supersaturation would not occur by surface nucleation followed by spreading of two-dimensional layers. Frank and his collaborators showed that under such conditions, dislocations would be the controlling factor in crystal growth. Frank [1] pointed out that the emergence of a screw dislocation on a crystal face produces on that face a step of height equal to that of the corresponding Burgers vector. A principal crystal face that contains the end of screw dislocation will appear essentially like the one shown in Figure 5.1. It will have a step whose termination is not at the boundary but at the emergence point of the dislocation. The growth of the crystal on such a surface does not eliminate the step because if we assume that the crystal is growing by the attachment of molecules to the edge of this step, the step in question would be self-perpetuating and continues to be present on the crystal surface as long as the dislocation line intersects the surface. Since the step provided by the screw dislocation terminates at the dislocation line, it can advance only by rotating around the dislocation. Hence, the step winds itself into a closed spiral centered on the dislocation, and as the growth proceeds the spiral revolves around, as shown sequentially in Figure 5.1. This will also be true for any dislocation whose Burgers vector has a component normal to the face in which the dislocation terminates. Unlike the ideal crystals of "n" layers, the crystal containing a screw dislocation consists of only one layer in the form of a helicoid. When the dislocation is of multiple strengths, we have a crystal of several interleaved helicoidal layers.

In general, the emergence of screw dislocations in a crystal face eliminates the need for surface nucleation, provided the distance between pairs of dislocations or between the dislocation and the edge of the crystal exceeds the diameter of the critical surface nucleus for the supersaturation imposed. Such crystals can grow at almost arbitrarily low values of the supersaturation. A crystal whose faces are only twice as wide as the critical nucleus will be the limiting case, which grows if it contains just one screw dislocation at the center.

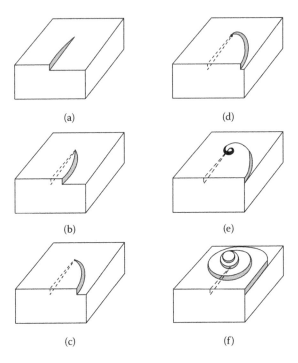

FIGURE 5.1
Crystal growth due to the presence of a screw dislocation. (a–f) progressively show how the step winds itself to form a spiral.

The Theory

We can now derive an expression for the growth rate of a crystal face containing a screw dislocation and infer its dependence on supersaturation.

Earlier we saw that the rate of advance of a straight step and that of a curved step with radius of curvature ρ are given by

$$v_\infty = 2\,(\alpha - 1)\,x_s\,\nu\,e^{-W/kT} \tag{5.1}$$

$$v_\rho = v_\infty\,(1 - \rho_c/\rho) \tag{5.2}$$

where $\rho_c = a\varphi/2kT\,log\,\alpha$.

Hence, the step created by a screw dislocation initially advances straight parallel to itself with constant velocity v_∞ over the greater part of its length. However, in the neighborhood of dislocation where such advance would produce a sharp corner, it advances more slowly in such a way that its curvature is everywhere less than that corresponding to the critical radius ρ_c^{-1}. This causes the step to wind up into a spiral centered on the dislocation end with approximately constant spacing between the turns. The three cases of step advancement are schematically shown in Figure 5.2.

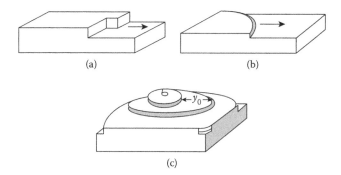

FIGURE 5.2
Advancement of a monolayer with (a) a straight step, (b) curved step, and (c) sequence of curved parallel steps.

Burton and Cabrera [2], after determining the rate of advance of a straight step, have extended their work to derive an expression for the rate of advance of parallel sequence of steps separated by equal distance y_0 [3]. The rate of advancement of parallel sequence of steps is given by the expression

$$v_\infty = 2\,(\alpha - 1)\,x_s\,v\,e^{-W/kT}\,\tanh\,(y_0/2x_s) \tag{5.3}$$

Now, the angular velocity of the spiral is given by

$$\omega = \frac{v_\infty}{2\rho_c} \tag{5.4}$$

Hence, the number of turns of spiral per second that passes through any point is $\omega/2\pi$. This is called the activity of the spiral.

Assuming $y_0 = 4\pi\rho_c$ (justified by the experimental results), the actual rate of vertical growth of the pyramid and therefore that of the crystal face will be

$$\Gamma = \frac{\omega d}{2\pi} = \frac{v_\infty d}{4\pi\rho_c} \tag{5.5}$$

where d is the step height. Substituting Equation 5.3 in Equation 5.5, we get

$$\Gamma = \frac{2(\alpha-1)x_s d v e^{-W/kT}}{4\pi\rho_c}\tanh\frac{4\pi\rho_c}{2x_s} = \frac{(\alpha-1)d v e^{-W/kT}}{\dfrac{2\pi\rho_c}{x_s}}\tanh\frac{2\pi\rho_c}{x_s} \tag{5.6}$$

Now, let $\dfrac{2\pi\rho_c}{x_s} = \dfrac{\beta}{\alpha-1}$

This is consistent with the fact that for small values of α, ρ_c can be approximated to

$$\rho_c \approx \frac{a\varphi}{2kT(\alpha-1)} \tag{5.7}$$

Then,

$$\Gamma = \frac{(\alpha-1)^2}{\beta}dve^{-W/kT}\tanh\frac{\beta}{\alpha-1} \tag{5.8}$$

Now for low supersaturation, $(\alpha-1) \ll \beta$ and hence $\tanh\dfrac{\beta}{\alpha-1} = 1$

or

$$\Gamma = \frac{(\alpha-1)^2}{\beta}dve^{-W/kT} \tag{5.9}$$

a parabolic law with respect to supersaturation.

For high supersaturation $(\alpha-1) \gg \beta$ and hence $\tanh\dfrac{\beta}{\alpha-1} = \dfrac{\beta}{\alpha-1}$

or

$$\Gamma = (\alpha-1)dve^{-W/kT} \tag{5.10}$$

a linear law with respect to supersaturation.

We can see that there is a critical supersaturation given by the formula for β below which the rate of growth is parabolic and above which it is linear.

Similar considerations apply to the rate of growth produced by a group of dislocations. If the group contains an equal number of right- and left-handed dislocations (i.e., zero total strength), then there must be another critical supersaturation β_ℓ below which no growth will occur. β_ℓ will be defined by the condition $2\rho_c = \ell$, where ℓ is the maximum distance between pairs of dislocations actually coupled by step.

Experimental Verification

One can theoretically estimate the growth rate dependence on supersaturation using the aforementioned expressions. Crude theoretical estimates of v and x_S for iodine at 0°C give $vd \sim 4kT$ and $x_s \sim 400a$. The resulting theoretical curve was found to be in reasonably good agreement with the observations of Volmer and Schultze [4].

The number of experimental observations in support of the screw dislocation theory has increased quite considerably since the development of screw dislocation theory of crystal growth. A collected account of early observations are given by Buckley [5], Verma [6], Forty [7], and Amelincks [8]. Micro-topographical studies of crystal faces have revealed the presence of growth spirals on the habit faces of a large number of crystals. Optical micrography,

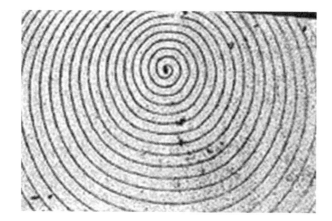

FIGURE 5.3
Typical growth spiral. (From Verma, A. R., and Amelinckx, S., *Nature*, 167, 939–940, 1951. With permission.)

coupled with multiple-beam interferometry developed by Tolansky [9], has helped the researchers to identify growth spirals with a great degree of certainty. From the fringe shift observed in multiple-beam interferograms taken on crystal surfaces having growth spirals, one could calculate the step height of the spirals. This compared well with the d spacing (or $n \times d$, where n is an integer) of the plane on which it was observed. A typical growth spiral is shown in Figure 5.3 [10].

Growth from Solution

Crystal growth theories developed initially for growth from vapor phase can be applied to growth from solution. Although it is clear from the qualitative point of view that there is no essential difference between the growth from vapor and from solution, a quantitative theory establishing the dependence of the rate of growth on supersaturation in the case of growth from solution is much more difficult. Burton et al. [3] treated the problem in the following way.

First of all, in the case of growth from solution, we should expect the rate of advance of a step in the crystal surface to be slow because the diffusion of solute molecules toward the kinks, through the solution either on the surface or in the edge of the step, is now much slower than in the free surface of a crystal. It is difficult to decide the relative importance of these diffusion currents. However, in order to simplify the problem, one can neglect the contributions from the diffusion on the surface and in the edge and consider only diffusion in the bulk. Even under these conditions, neglecting the thermal motion of kinks is an oversimplification of the problem and hence really not justified.

Let us imagine that we have a set of parallel steps at distances y_0 from each other on a crystal surface in one of the close-packed crystallographic directions (for which x_0, the mean distance between the kinks, is a maximum). As an approximation, the diffusion through the solution can be broken up into three components as follows:

1. At distances $r < x_0$ from each kink, we have a hemispherical diffusion field around each of them with the diffusion potential

$$\left(1 - \frac{a}{r}\right)\alpha(x_0) \tag{5.11}$$

   where $\alpha(x_0)$ is the supersaturation at a distance r from the kink.

2. At distances r between x_0 and y_0 from each step, we have a semicylindrical diffusion field around each step with diffusion potential

$$\left[\log\frac{y_0}{x_0}\right]^{-1}\left[\alpha(x_0)\log\frac{y_0}{r} + \alpha(y_0)\log\left(\frac{r}{x_0}\right)\right] \tag{5.12}$$

   where $\alpha(y_0)$ is the supersaturation at y_0.

3. Finally, at distances z from the crystal surface between y_0 and δ, the thickness of the unstirred layer at the surface of the crystal, we have a plane diffusion field with the diffusion potential

$$\left[(z - y_0)\alpha(z) + (\delta - z)\alpha(y_0)\right](\delta - y_0)^{-1} \tag{5.13}$$

   where $\alpha(z)$ is the supersaturation in the stirred solution. The latter diffusion potential applies to any case, but we cannot equate the flux calculated from its gradient to the rate of growth R of the crystal unless $D/R \gg \delta$, where D is the diffusion coefficient in the solution.

Equating the sum of the hemispherical fluxes going to all the kinks in the step to the semicylindrical flux, and equating the sum of the semicylindrical fluxes going to all the steps in the surface to the plane flux, Burton et al. eliminated $\alpha(y_0)$. The resulting equation yielded a parabolic law at low supersaturation (below ~10^{-3}) and linear law at high supersaturation. This is known as the volume diffusion model of Burton et al. [3].

In subsequent years, several attempts were made to improve our understanding of crystal growth from solutions, leading to the development of models such as the volume diffusion model of Chernov [11], the combined volume and surface diffusion model [12], the dehydration model [13], the surface reaction model [14,15], and so on. A review of these developments has been given by Konak [16]. However, he concludes from his critical review that "no theory has yet been proposed to give a complete picture of the growth process" and the situation has remained more or less the same even today.

Shape of Growth Spirals

In the foregoing discussion, we have assumed a circular symmetry for the growth spiral (a paradoxical phrase). This would not be the case if the growth rate varies with orientation in the crystal face. An obvious source of variation is when the step is parallel to certain close-packed directions. Such a step may be relatively free from kinks because the kink energy ω would then be high. Hence, the rate of advancement of step in this direction will be reduced resulting in a polygonal spiral.

The rotational symmetry of a polygonal growth spiral must possess at least the rotational symmetry of the crystal face on which it appears, assuming a symmetrical environment. Figure 5.4 [17] shows such a spiral observed on tungsten selenide crystal grown by sublimation process. It may be noted that the center of the spiral appears to be circular. However, soon the crystal symmetry takes over and polygonal symmetry results. The segments of the polygons will be straight only in the limit of very low supersaturation. In general, they will be curved convexly (Figure 5.5) [18], but the corners will have essentially the same curvature as that of a critical nucleus.

Since the spacing between successive turns of a growth spiral is proportional to the diameter of the critical nucleus, it is essentially inversely proportional to the supersaturation. A circular or polygonal spiral on the crystal face will appear macroscopically as a pyramid of vicinal faces, the slope of which is inversely proportional to the spacing between turns and therefore proportional to the supersaturation. Since the height of the spiral steps is very small, the vicinal faces make a very small angle with respect to the low-index planes on which the spiral appears.

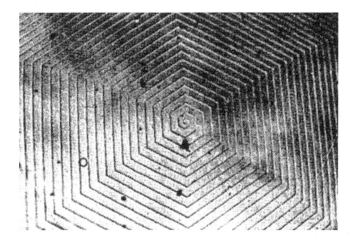

FIGURE 5.4
Polygonal growth spiral observed in a tungsten selenide crystal. (From *Journal of Crystal Growth*, 41, Agarwal, M. K., et al., Growth of single crystals of WSe_2 by sublimation method, 84–86, 1977, with permission from Elsevier.)

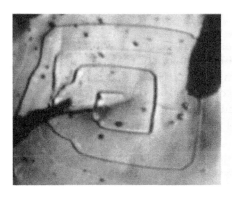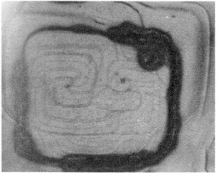

FIGURE 5.5
Polygonal growth spiral showing curved segments. (Reprinted from *Journal of Crystal Growth*, 121, Elizabeth, S. G., et al., Growth and extraction of flux free YBCO crystals, 531–535, 1992, with permission from Elsevier.)

Interaction of Growth Spirals

In the theory presented earlier only one dislocation is assumed to be present. The presence of more dislocations makes practically little difference. The detailed theory of interactions between growth spirals is discussed by Burton et al. [3]. Two dislocations separated by a distance larger than $2\pi\rho_c$ can act as centers of two regions whose growth steps proceed from them. The growth rate is everywhere the same as if only one dislocation were present, but the examination of vicinal faces will reveal two growth hillocks. However, if the supersaturation at the center of one is slightly greater than that of the other, the territory of the second will continuously shrink; and if the conditions remain constant, then only one growth hillock will be visible. This is a case of one dislocation dominating the other. If two dislocations of the same sense are closer together than $2\pi\rho_c$, they generate a pair of nonintersecting growth spirals rather than behaving like a dislocation of double strengths. They are now said to be cooperating and send out growth steps twice as fast as single dislocation. The growth rate in regions where parabolic law holds good is thus doubled. This implies that the regime of linear law would be reached sooner.

Vapor–Liquid–Solid Mechanism of Crystal Growth

Although screw dislocation theory was highly successful in explaining the crystal growth under low supersaturation conditions, there were experimental observations that cast doubt on several aspects of this theory of crystal growth. The step heights of growth spirals were often too large to be understood in terms of screw dislocation and their developments on the slow-growing faces. Further, studies on whisker growth from vapor raised the doubt about the probable role of screw dislocation theory in their growth. Scientists were looking for an alternative explanation.

Inhibition of growth in two dimensions gives rise to what are called whisker crystals. A whisker is a long filamentary crystal with cross-section (diameter) less than 100 μm (sometimes very much less) but lengths may extend to centimeters. Figure 5.6 shows a large number of gallium phosphide (GaP)

(a)

(b)

20μ

FIGURE 5.6
Gallium phosphide (GaP) whiskers grown using H_2O-Ga-GaP system: (a) inner surface of the growth tube. (Schönherr, E., *Journal of Crystal Growth*, 9, 346–350, 1971. With permission.); (b) straight, bent, and coiled whiskers, bending presumably due to convection-driven instability. (Schönherr, E., and Winckler, E., *Journal of Crystal Growth*, 32, 117–122, 1976. With permission.)

whiskers grown using H₂O-Ga-GaP system [19,20]. One of the characteristic properties of whisker crystal is its great mechanical strength, which is attributed to its high structural perfection. Why the sides of the whisker do not grow at appreciable rates is not precisely understood. With very anisotropic substance, this can probably be explained by the existence of an easy growth direction and lateral faces on which nucleation is very difficult. However, many cubic materials grow in whisker form and hence require a different explanation.

The explanation given on the basis of screw dislocation theory is that in a whisker crystal there are dislocations only along the whisker axis, while in the other two dimensions the faces will be perfect. Consequently, no growth will occur at an appreciable rate on these faces (the side faces of the whiskers are believed to grow by the two-dimensional layer growth mechanism), but growth occurs along the whisker axis due to the presence of axial screw dislocations. The cause for the presence of dislocations in certain direction is not exactly known. It has, however, been suggested that their presence might be due to the nature of substrate on which the nucleation has occurred. In particular, if substrate contained a screw dislocation, then this might be transmitted to the whisker.

A search for the axial screw dislocation in crystals grown in the form of whisker has given surprising results. A large number of cases have been reported in which whiskers have been grown without axial screw dislocation at all. This led Wagner and Ellis [21–23] to put forward their vapor–liquid–solid (VLS) mechanism of growth. The VLS mechanism of growth can be best illustrated by the schematic diagram, which describes the method used in Si whisker growth (Figure 5.7). Deposition from vapor

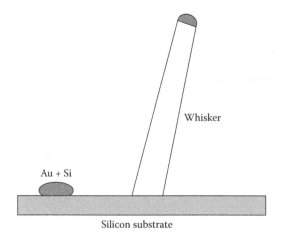

FIGURE 5.7
Schematic of the VLS mechanism.

is either by iodide disproportionation or by hydrogen reduction of $SiCl_4$, as per the following reactions.

$$2\ SiI_{2(s)} \rightarrow SiI_{4(g)} + Si_{(soln)}$$

$$SiCl_{4(g)} + 2\ H_{2(g)} \rightarrow 4\ HCl_{(g)} + Si_{(soln)}$$

The liquid phase is a solution of Si in gold. In the process of growth, the vapor phase reaction does not directly result in solid Si. Instead, a liquid solution of Si in gold is formed. The solid Si then grows from the liquid solution.

$$Si_{soln} \rightarrow Si_{solid}$$

The deposition from liquid is aided by the fact that a large thermal gradient at the growing interface is imposed. The initial growth mechanism is obscured somewhat, but the growth is easy to initiate. In a typical growth experiment, a small particle of gold is placed on a clean (111) face of single crystal Si wafer and heated to 950°C, where it forms a small liquid Au–Si alloy droplet (Figure 5.7). Then, the required vapors are introduced and the growth of whisker begins. As the growth progresses, the alloy droplet rides atop the whisker. Usually, a small concentration of the Au remains behind in the grown whisker as a globule. However, if the whisker is sufficiently long, all the Au gets consumed. The grown crystal is dislocation-free. Indeed, as more VLS investigations were undertaken, it became clear that most whiskers grew by VLS mechanism and not by screw dislocation mechanism. The subsequent vapor growth studies have revealed that the VLS mechanism is also important for the growth of platelets, films, and even bulk crystals, although it is whiskers that indicated the new mechanism.

The discovery of the VLS mechanism and the investigations that followed have generated great interest in liquid layers on vapor-grown crystals. Givargizov [24] has dealt with various aspects of whisker growth in his review paper.

Dendritic Growth

One of the commonest modes of growth by which crystals of every possible crystal symmetry and chemical composition grow is by the formation of dendrites. They are almost invariably a rule in metals and are also observed in crystals grown from melt, solution, and vapor.

A dendrite is a crystal with a tree-like branching structure. A typical dendrite consists of a primary stem on which secondary branches grow with tertiary branching occurring on the latter (Figure 5.8). This commonly occurs in one plane, but three-dimensional dendrites have also been observed [25,26].

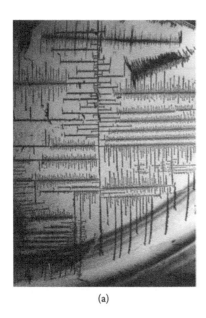
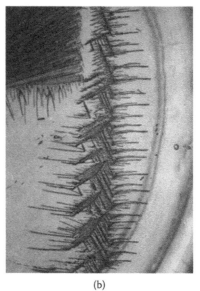

(a)

(b)

FIGURE 5.8
Dendritic crystals of lead chloride. (a) Two dimensional and (b) three dimensional (3x). (With kind permission from Springer: *Journal of Materials Science*, Filamentary and dendritic growth of lead chloride crystals in silica gel, 16, 1981, 1707–10, Bhat, H. L.)

Dendrites are normally single crystals, and the branches follow definite crystallographic directions. The branches are often regularly spaced, and the opposite sides of the primary stem show marked symmetry. At times, dendrites whose limbs are not related to any crystallographic directions are also observed. In later stages, however, the dendrites often overlap and fill in, and the filling-in process is indistinguishable from the processes that produce uniform crystals. Even though the cause of this abnormal growth is high degree of supersaturation, no definitive explanation regarding the regularity of dendrite branching is available. However, the spacing of the branches depends on the supersaturation of the solution. Since the growth is occurring under a high supersaturation condition, it is unlikely that dislocations play any significant role in their growth. Early theories on dendritic growth have been given by Buckley [5].

References

1. Frank, F. C. 1949. The influence of dislocations on crystal growth. *Discussions of the Faraday Society* 5: 48–54.
2. Burton, W. K., and N. Cabrera. 1949. Crystal growth and surface structure. Part I. *Discussions of the Faraday Society* 5: 33–39.

3. Burton, W. K., N. Cabrera, and F. C. Frank. 1951. The growth of crystals and the equilibrium structure of their surfaces. *Philosophical Transactions of the Royal Society of London* 243: 299–358.

4. Volmer, M., and W. Schultze. 1931. Condensation on crystals. *Zeitschrift fur Physikalische Chemie* 156: 1–22.

5. Buckley, H. E. 1951. *Crystal growth*. New York, NY: John Wiley.

6. Verma, A. R. 1953. *Crystal growth and dislocations*. London, UK: Butterworth Publications.

7. Forty, A. J. 1954. Direct observations of dislocations in crystals. *Advances in Physics* 3: 1–25.

8. Amelincks, S. 1954. Growth spirals of crystals of long chain compounds. *Naturwissenschaften* 41: 356–357.

9. Tolansky, S. 1948. *Multiple-beam interferometry of surfaces and films*. Oxford, UK: Oxford University Press.

10. Verma, A. R., and S. Amelinckx. 1951. Spiral growth on carborundum crystal faces. *Nature* 167: 939–940.

11. Chernov, A. A. 1961. The spiral growth of crystals. *Soviet Physics Uspekhi* 4: 116–145.

12. Gilmer, G. H., R. Ghez, and N. Cabrera. 1971. An analysis of combined surface and volume diffusion processes in crystal growth. *Journal of Crystal Growth* 8: 79–93.

13. Von Reich, R., and M. Kahlweit. 1968. Zur kinetic les Kristallwachstums in wassrigen losungen I-II. *Berichte der Bunsengesellschaft fur Physikalische Chemie* 72: 66–74.

14. Walton, A. G. 1963. A theory of surface reaction controlled growth. *Journal of Physical Chemistry* 67: 1920–1922.

15. Doremus, R. H. Crystallization of slightly soluble salts from solution. *Journal of Physical Chemistry* 74: 1405–1408.

16. Konak, A. R. 1973. Difficulties associated with theories of growth from solutions. *Journal of Crystal Growth* 19: 247–252.

17. Agarwal, M. K., H. B. Patel, and K. Nagireddy. 1977. Growth of single crystals of WSe_2 by sublimation method. *Journal of Crystal Growth* 41: 84–86.

18. Elizabeth, S., G. Dhanaraj, H. L. Bhat, and S. V. Bhat. 1992. Growth and extraction of flux free YBCO crystals. *Journal of Crystal Growth* 121(3): 531–535.

19. Schönherr, E. 1917. Photographic observation of the growth of GaP-needles from the wet hydrogen process. *Journal of Crystal Growth* 9: 346–350.

20. Schönherr, E., and Winckler E. 1976. Bending and straightening of GaP whiskers during their growth. *Journal of Crystal Growth* 32: 117–122.

21. Wagner, R. S., and W. C. Ellis. 1964. Vapor-liquid-solid mechanism of single crystal growth. *Applied Physics Letters* 4: 89–90.

22. Wagner, R. S., and W. C. Ellis. 1965. The vapor-liquid-solid mechanism of crystal growth. *Transactions of the Metallurgical Society of AIME* 233: 1053–1064.

23. Wagner, R. S. 1970. VLS mechanism of crystal growth. In *Whisker technology*, edited by A. P. Levitt. New York, NY: John Wiley, pp. 47–119.

24. Givargizov, E. I. 1975. Fundamental aspects of VLS growth. *Journal of Crystal Growth* 31: 20–30.

25. Chalmers, B. 1964. *Principles of solidification*. New York, NY: John Wiley, p. 93.

26. Bhat, H. L. 1981. Filamentary and dendritic growth of lead chloride crystals in silica gel. *Journal of Materials Science* 16: 1707–1710.

6

Phase Diagrams and Crystal Growth

In a broad sense, crystal growth can be regarded as the science and technology of controlling phase transitions that lead to crystalline solids. The processes involved in crystal growth are traditionally referred to as sublimation, solidification, precipitation, recrystallization, and so on. The growth as such involves a first-order phase transition characterized by a jump in the heat content, volume, and several other physical properties. In addition, the crystal might undergo one or more phase transitions when it is being cooled down from growth temperature to room temperature, and often such transitions would greatly influence the structural perfection of the grown crystals. Well-known α–β transition of quartz and ferroelectric phase transitions in $BaTiO_3$, $LiNbO_3$, and $KTiOPO_4$ are examples of such transitions that occur when the grown crystal cools down to room temperature. In principle, the phase diagrams would include such information as well. Furthermore, materials may melt congruently or incongruently. During congruent melting, the composition of the liquid that forms is the same as that of the solid. In contrast, if a compound melts incongruently, the composition of the resulting liquid will not be the same as that of the solid and one has to use a phase diagram to determine it. Consequently, knowledge of the phase diagrams is one of the essential inputs that help in deciding the method to be employed to grow a particular crystal.

In this chapter, we first briefly introduce the thermodynamics of phase transitions. However, the treatment of the subject is not comprehensive, and only such aspects that are relevant to crystal growth are discussed. Some practical phase diagrams are described to emphasize the importance of phase diagrams to crystal growth. Excellent books written on phase diagrams should be consulted for a deeper understanding of the subject [1–3].

Latent Heat

Since a phase transition of first order involves latent heat, we must consider it as an important thermodynamic quantity that has relevance to crystal growth. In particular, when a crystal is being grown from the melt at a relatively high growth rate, the released latent heat of fusion would be significant and hence bound to influence its growth. It is to be noted that latent heats vary widely, covering two orders of magnitude. In general, one should keep in mind the following relationships between various kinds of latent heat while attempting to grow crystals from the three phases, that is, solid, liquid, and vapor.

$$\Delta H_{vap} > \Delta H_{fus} > \Delta H_{s/s},$$

$$\Delta H_{sub} = \Delta H_{fus} + \Delta H_{vap}, \text{ and } \Delta H_{sub} \gg \Delta H_{fus}$$

Heat of fusion has profound influence on the shape of the phase diagram. Muhlberg [4] has shown how heat of fusion of the individual components influences the phase diagrams. His calculation using the van Laar equation (the van Laar equation was developed by Johannes van Laar in 1910–1913 to describe phase equilibria of liquid mixtures) has shown that for an ideal system, the higher the heat of fusion, the broader the width between liquidus and solidus lines. Furthermore, he shows that the difference between the heats of fusion determines the asymmetric shape of the phase diagram. The consequence of the shape of the phase diagrams on the segregation behavior in normal freezing growth processes will be discussed in Chapter 8.

Gibbs's Free Energy

Another important thermodynamic quantity relevant to crystal growth processes is, of course, the Gibbs's free energy G, which is the convertible energy due to phase change. In a single-component system, when the two phases (solid and melt or liquid or vapor) are in equilibrium, one can show that

$$dG_{ph1} = dG_{ph2}$$

or

$$(V^{ph2} - V^{ph1})dp - (S^{ph2} - S^{ph1})\, dT = 0$$

Since $(S^{ph2} - S^{ph1}) = \left(\dfrac{\Delta H}{T}\right)_{phase\ trans}$ we can write, $dV\, dp - \Delta H \dfrac{dT}{T} = 0$

Hence,

$$\frac{dp}{dT} = \frac{\Delta H_{tr}}{T\Delta V} \quad \text{or} \quad \frac{\Delta T}{dp} = \frac{T\Delta V}{\Delta H_{tr}} \qquad (6.1)$$

This is the well-known Clausius–Clapeyron equation. For our case, it describes the temperature dependence of vapor pressure and the dependence of melting temperature on vapor pressure, respectively. Thus, in principle by imposing an external pressure, one can increase the melting temperature of a system. However, it has been generally found that large pressure changes are required to shift the melting point of materials just by a few degrees. This makes pressure control rather impractical for exploiting it to improve the operating conditions of crystal growth process. Nevertheless, high-pressure crystal growth has been employed to elevate the freezing point of certain materials. For example, Harrison and Tiller [5] have used high-pressure growth to obtain single crystals of hexagonal selenium from melt by raising the melting point. Application of pressure also had the advantage of lowering the viscosity of the selenium melt, thereby augmenting the rate of crystal growth.

Chemical Potential

An equally important thermodynamic quantity, but rather difficult to get a real feeling of, is the chemical potential μ. Chemical potential can be understood in terms of Gibbs's free energy per mole of a substance and is also sometimes referred to as partial Gibbs's molar free energy. An intuitive way to conceptualize chemical potential is to realize that every substance has a tendency to change; that is, to react with other substances, to transform into another state of aggregation, or to move to another place. This tendency can be described by a single quantity called the chemical potential. For a given substance, the chemical potential is unique, although it changes with temperature, pressure, and concentration. Systems always tend to move from higher chemical potential to lower chemical potential.

In thermodynamics, chemical potential is a form of potential energy that can be absorbed or released during such a change. Hence, it is an important parameter in processes involved in crystal growth such as melting, boiling, evaporation, sublimation, solubility, and so on. For a system containing n constituent species with ith species having N_i particles, the fundamental thermodynamic equation can be written in terms of Gibbs's free energy as

$$dG = -SdT + Vdp + \sum_{i}^{n} \mu_i dN_i \qquad (6.2)$$

At constant temperature and pressure, this equation simplifies to

$$dG = \sum_{i}^{n} \mu_i dN_i = \mu_1 dN_1 + \mu_2 dN_2 + \dots\dots \qquad (6.3)$$

The definition of chemical potential μ_i of the ith species can be written by setting all the numbers N apart from the one under consideration constant. Thus

$$\mu_i = \left(\frac{\partial G}{\partial N_i} \right) T, P, N \quad j \neq i \tag{6.4}$$

It must be noted that it is only with first-order phase transition that at the transition point, two thermodynamically distinct phases are in equilibrium with each other (i.e., $\mu_2 = \mu_1$). For second-order phase transitions, this is not the case. For example, at critical temperature T_c of a second-order (order–disorder) transition, one cannot distinguish between an ordered and disordered phase.

Gibbs's Phase Rule

The main key to understanding phase diagrams is the Gibbs's phase rule developed in 1870 [6] and given by

$$P + F = C + 2 \tag{6.5}$$

where P is the number of phases, C is the number of components in the system, and F is the number of degrees of freedom (also known as variance). If pressure is excluded as a variable, which would imply that the process is taking place at constant pressure, the number of independent variables will be reduced to two and we will then have a condensed phase rule given by

$$P + F = C + 1 \tag{6.6}$$

The basis of the rule is that the equilibrium between the phases introduces a constraint on the intensive variables such as concentration, temperature, pressure, specific volume, and so on. Since the phases are in thermodynamic equilibrium with each other, the chemical potentials of the phases must be equal. The number of degrees of freedom is then determined by the number of such equality relationships. For example, if the chemical potentials of liquid and its vapor depend on the temperature and pressure, the equality of chemical potential will mean that each of these variables will depend on each other; that is,

$$\mu_{liq}(T, p) = \mu_{vap}(T, p) \tag{6.7}$$

For a single-component system $C = 1$, so that we can write the phase rule as $F = 3 - P$. For such a system in a single phase ($P = 1$) region, two variables ($F = 2$), such as temperature and pressure, can be varied independently.

However, if the values are such that they lead to phase separation ($P = 2$), F decreases from 2 to 1. Hence, it would not be possible to independently vary temperature and pressure. Also, for a single-component system, it is possible that three phases, such as solid, liquid, and vapor, coexist in equilibrium ($P = 3$). Then, there are no degrees of freedom ($F = 0$). Consequently, this three-phase mixture can only exist at a single temperature and pressure, which is known as a triple point. Here,

$$\mu_{\text{sol}}(T, p) = \mu_{\text{liq}}(T, p) = \mu_{\text{vap}}(T, p) \tag{6.8}$$

Using this, the two variables T and p can be determined.

If four phases of a single-component system were to be in equilibrium ($P = 4$), the phase rule would give $F = -1$, implying -1 independent variables. This has no meaning and explains why four phases of a pure substance are not found in practice to be in equilibrium at any temperature and pressure. In terms of chemical potentials there are now three equations, which cannot in general be satisfied by any values of the two variables T and p. In practice, however, the coexistence of more phases than allowed by the phase rule normally means that the different phases are not in true equilibrium.

For a two-component system, the maximum number of phases at equilibrium is four, as per the phase rule, at which point the system has zero degree of freedom.

Phase Diagrams

Phase diagrams actually depict the thermodynamic laws and rules between various phases existing in a pressure–temperature–composition (p–T–x) phase space. They are classified by the number of components present in a system as one-, two-, and three-component systems. Single-component systems are represented by unary phase diagrams. Two-component systems have binary phase diagrams, three-component systems are represented by ternary diagrams, and so on. Where more than two components are present, the phase diagrams become quite complicated and will be difficult to represent, but we can give their binary sections.

Single-Component Systems

There are many single-component systems such as C, SiO_2 (which is the starting material to grow quartz crystals), ZrO_2, Fe, Al_2SiO_5, and so on, which are of interest to crystal growers. We will start with the classic single-component system (i.e., water) and then take up a more complex system.

Water

The classic example of a one-component system is water, and a schematic phase diagram of it is shown in Figure 6.1. In this system, since there is no composition change ($C = 1$), the only variables are temperature and pressure. There are three distinct areas where only ice, water, or vapor exists, respectively. These are divariant fields because in these regions the two variables can be varied independently, and we will still have only one phase. Along the phase boundary, two phases coexist and hence only temperature or pressure can be varied independently if we want both the phases to coexist. Such a boundary is called univariant curve. Also, we have a unique point ($T = 0.01°C$ and $P = 0.006$ atm) where all the three phases (i.e., ice, water, and water vapor) coexist, and this is, as mentioned earlier, an invariant point called the triple point. The end of the curve (boiling curve) separating the liquid from vapor transition is called the "critical point." This part of the phase diagram is particularly interesting because beyond the critical point the physicochemical properties of water and steam converge to become identical, and we have a single phase. We refer to this single phase as a "supercritical fluid." Please note that in the case of water, the slope of the ice–water phase boundary is negative; that is, the volume of ice is greater than that of water for the same mass. Such a volume change poses problems during crystal growth, particularly when a crystal grows in contact with the crucible. However, in most cases the volume of liquid phase is greater than that of the solid phase.

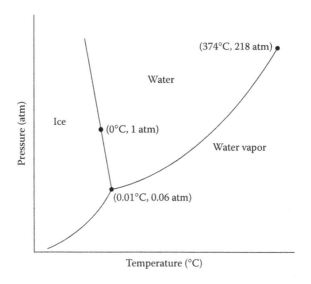

FIGURE 6.1
Phase diagram of water. Drawing not to scale.

Diamond

One of the interesting single-component systems that are relevant to crystal growers is diamond. Diamond is composed of the single element carbon, and it is the tetrahedral arrangement of the C atoms in the lattice that gives diamond its unique properties. While diamond is the hardest known natural material, graphite, which is an allotropic form of carbon, is one of the softest, simply because of the way the C atoms are bonded together in it. While in diamond each carbon is bonded to four others in a tetrahedral configuration, in graphite each carbon atom is bonded only to three others in sheets of connected benzene rings. Since the sheets interact through Van der Waal's forces, they can slide over one another; hence, graphite is soft and works as a good high temperature lubricant. As can be seen in the phase diagram of carbon shown in Figure 6.2 [7,8], there is a metastable region around the line separating diamond and graphite. In this region, the relationship between diamond and graphite is both a thermodynamic and kinetic one. At normal

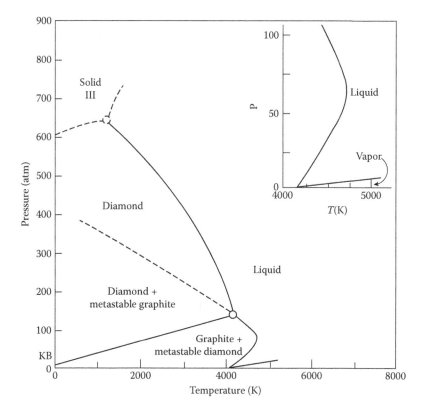

FIGURE 6.2

Phase diagram of carbon. (Reprinted with permission from Steinbeck, J, et al., A model for pulsed laser melting of graphite, *Journal of Applied Physics*, 58, 4374–4382, 1985, American Institute of Physics.)

temperatures and pressures, graphite is only slightly more stable than diamond (by a few electron volts). The fact that diamond exists is due to the very large activation barrier for conversion to graphite. Consequently, once diamond is formed, it cannot revert back to graphite. Therefore, diamond is said to be metastable, even though it is kinetically stable but not thermodynamically so. Diamond is understood to have been created deep underground under conditions of extreme pressure and temperature. Under these conditions, diamond is actually more stable than graphite; hence, over a period of millions of years, carbonaceous deposits slowly crystallize into single crystals of diamond. As you will see in Chapter 10, synthetic diamonds are grown essentially by mimicking the nature.

Two-Component Systems

In representing the two-component systems, it is generally assumed that pressure is constant so that condensed Gibbs's phase rule is applicable. Hence, two-component systems are usually drawn in terms of variation in composition with respect to temperature, and such diagrams are usually called binary phase diagrams.

Binary systems are classified according to their solid solubility. In case of limited solid solubility, there would be solid state miscibility gaps. Also, in this case a number of invariant reactions might take place, intermediate phases may exist over a range of compositions, or a single intermediate phase might occur at a fixed composition to form a definite compound. All these will make the phase diagram more complicated. In the following, we discuss some typical (but hypothetical) phase diagrams and trace the crystallization history at certain arbitrary compositions.

Two-Component Solid Solutions

In nature, substitution of one element by another often occurs because these elements are similar chemically and size-wise. When such substitutions occur, the resulting phase can have a range of possible compositions, depending on the amount of substitution that takes place. Such solids that can accommodate elemental substitution by various amounts are called solid solutions. When all compositions between two end members are possible, the solid solution is said to be a complete solid solution. If the elements that substitute are not exactly the same size (but similar in size), the amount of substitution depends on temperature and pressure.

Figure 6.3 shows a hypothetical binary phase diagram of a complete solid solution in which the two constituents are A and B. The upper curve is called the liquidus and the lower curve the solidus. At temperatures above the liquidus, the system is in the liquid state, and below the solidus, it is in the solid (crystalline) state. At temperatures between the solidus and liquidus, crystals of solid solution AB coexist with the liquid in equilibrium.

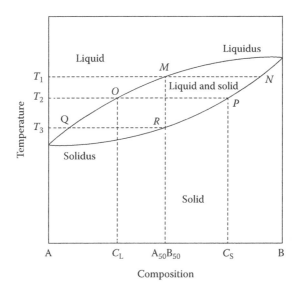

FIGURE 6.3
Simple binary phase diagram of a solid solution.

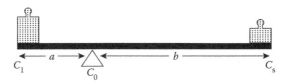

FIGURE 6.4
Pictorial representation of the lever rule.

Lever Rule

To specify completely the two phases at any temperature in a binary, both the composition of each phase at that temperature and the relative quantities of the two phases must be known. The compositions can be directly read from the composition axis of the phase diagram. The quantities of each phase present may also be determined with the help of a rule called the lever rule. The lever rule can be explained by considering a simple balance (Figure 6.4). Let the composition C_0 of the alloy at a given point in the phase diagram be represented by the fulcrum, and the compositions of the two phases (phases A and B) by the ends of a bar. Here, the bar represents the tie line in the phase diagram passing through the considered point, which intersects the phase boundary lines on either side at compositions C_1 and C_s. The proportions of the phases present are then determined by the weights needed to balance the system.

Thus,

the fraction of phase A = $(C_s - C_0)/(C_s - C_l) = b/(a + b)$ and
the fraction of phase B = $(C_0 - C_l)/(C_s - C_l) = a/(a + b)$

The fractions can also be represented in percentage weight by multiplying them by 100.

Crystallization Path

Let us now trace the crystallization path of composition $A_{50}B_{50}$ in Figure 6.3 when it is being cooled. For this composition, the binary is completely liquid above T_1. Lowering of temperature down to the point M on the liquidus leads to the crystallization of a small amount of the solid solution. We assume that only one single crystal is growing on cooling. The composition of this crystal may be found by drawing a horizontal line (an isotherm) through the temperature T_1 and then by drawing a vertical line to the base of the diagram from the point where the isotherm intersects the solidus. The crystal that is in equilibrium with liquid will always have a higher percentage of B relative to the liquid. On lowering the temperature, crystallization progresses and the composition of the crystal changes along the solidus, by continuously reacting with the liquid, thereby getting enriched in A. Simultaneously, the composition of the liquid will change along the liquids, thereby also getting enriched in A. At a temperature T_2 the liquid composition will be as at point O, while the solid composition will be as at point P. The crystallization proceeds until the temperature reaches T_3, at which point the remaining liquid will have a composition corresponding to the point Q and the solid corresponding to the point R. By now, all of the liquid would have solidified and the final crystalline product will have the composition $A_{50}B_{50}$. During crystallization the amount of the solid continually increases while that of the liquid continually decreases and reaches zero. If at any point during the crystallization we wish to determine the amount of solid and liquid, the lever rule can be applied.

In the above discussion, we have assumed that equilibrium is maintained during the entire process of crystallization. In such a case, the crystal always remains in contact with the melt and crystallization is slow enough to allow continuous and complete reaction between the crystal and the melt. Under these conditions, the crystal will not change its composition beyond R and the final product is a homogeneous crystalline solid having the same composition as that of the initial melt. However, equilibrium crystallization is not practical.

In a more practical situation where the crystal is being continuously withdrawn from the melt, as in the case of crystal growth by the pulling technique, reaction of crystal with the melt is prevented. Under these conditions, the composition of the liquid will continuously change along the liquidus curve toward component A, its limiting composition being the pure A itself. Consequently, with continuous withdrawal of the growing crystal, the successively formed part of the crystal would become progressively rich in A and the final product would be pure A, although it would be a very small proportion of the initial amount. Thus, there would be smooth compositional gradient in the grown crystal.

If the crystal remains suspended in the liquid, but relatively rapid crystallization prevents complete reaction between the crystal and liquid (as in the case of the Kyropoulos technique), the effect will not be the same as above. The situation is equivalent to partial removal of the crystal from the melt, and the crystal in contact with the melt acts as a substrate on which solid increasingly rich in A crystallizes. The resulting crystal contains zones of differing composition, with the inner zones having more B and the outer zones having more A. The average composition of the zoned crystal, however, is that of the initial system.

Two-Component Eutectic Systems

Figure 6.5 shows the simplest of the two-component eutectic phase diagrams in which the compositions are plotted across the horizontal axis. Similar phase diagram is exhibited by Si–Au binary. The possible phases in this system are pure crystals of A, pure crystals of B, and liquid with compositions ranging between pure A and pure B. As shown in the phase diagram, the melting points of A and B are T_A and T_B, respectively. Figure 6.5 shows regions of liquid, A + liquid, B + liquid, and solid A + B separated

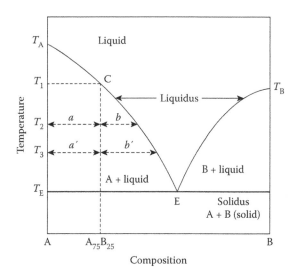

FIGURE 6.5
Binary phase diagram with a eutectic.

by respective liquidus and solidus curves. At the eutectic point E, where liquidus curves and the solidus intersect, all the three phases coexist and are in equilibrium.

If we assume pressure to be constant, Gibbs's phase rule for this system can be written as $F = C + 1 - P$. The eutectic point is an invariant point because here the degree of freedom is zero. Any change in either the composition of the liquid or the temperature will result in the number of phases reducing to two. One can notice that on adding B to pure A the melting temperature reduces monotonically, the lowest value being the eutectic temperature T_E. The same is true when A is added to B at the other end. Further, for all compositions between A and B, the melting also occurs over a range of temperatures between the solidus and the liquidus, except for the eutectic composition, which melts at T_E as in a pure solid.

Let us now consider the crystallization of a liquid with composition $A_{75}B_{25}$ as it is being cooled. Referring to Figure 6.5, composition $A_{75}B_{25}$ will be all liquid above T_1. If the temperature is lowered to T_1, crystals of A begin to form. Further lowering of the temperature causes more crystals of A to form. As a result, the liquid composition will continuously get enriched in B. As the temperature is lowered from T_1 to T_E through T_2 and T_3, the liquid composition will change from point C to point E. At all temperatures between T_1 and T_E, two phases (i.e., crystals of A and liquid mixture of A and B) will be present in the system. At the eutectic temperature, T_E, crystals of B will begin to form, and three phases—that is, crystals of A, crystals of B, and liquid mixture of A and B—will coexist. The temperature will remain at T_E until one of the phases disappears. Finally, when the liquid crystallizes completely, only pure solid A and pure solid B will remain, and the solid mixture will be in the proportions of the original mixture (75% A and 25% B).

If we were to stop the crystallization process at any temperature, say T_2, the amount of crystals of A and liquid could be determined by measuring the distances *a* and *b* on Figure 6.5. Their respective percentages can be calculated using the lever rule.

Percentage of crystals of A = $b/(a + b) \times 100$

Percentage of liquid = $a/(a + b) \times 100$

It must be noted that since the amount of crystals must increase with falling temperature, the proportional distance from the vertical line that marks the initial composition to the liquidus increases as temperature falls (see Figure 6.5). Thus, the distance used to calculate the amount of solid is always measured toward the liquid side of the initial composition and that used to calculate the amount of liquid is toward the solid side of the initial composition.

The melting process is exactly the reverse of the crystallization process. Thus, if we were to start with composition 75% A and 25% B at some temperature below T_E, the first liquid would form at T_E. The temperature would then remain constant at T_E until all of the crystals of B were melted.

As temperature increases, the liquid composition changes along the liquidus curve from point E to C until the temperature T_1 is reached. Above T_1, the system would contain only liquid with a composition of 75% A and 25% B.

Two-Component Peritectic Systems

It must be noted that not all binary systems whose components dissolve in each other form simple binaries or eutectics. Some of them have phase diagrams, which are classified as peritectic. Peritectics, such as eutectics, have extensive two-phase regions. They have unique compositions at which a change in temperature results in the transformation of a single phase into two phases, and this happens at a particular temperature called the peritectic temperature. Peritectic point, however, is the highest melting point of a particular solid phase. As we will see in Section II some of the technologically important crystals to be discussed in this book have peritectic phase diagrams.

Three-Component Systems

If instead of two, we have three components, say A, B, and C in a system, three binaries (i.e., AB, BC, and CA) may be formed as well as combinations of all the three elements. As per the Gibbs's phase rule, the state of a phase in a ternary system has four degrees of freedom. If we fix the pressure, the number of degrees of freedom reduces to three and hence we require three dimensions to represent an isobaric ternary phase diagram. Such a three-dimensional plot consists of a triangular-based prism with compositional variables plotted on the sides of the triangle and the temperature plotted on the vertical axis. The simplest ternary phase diagram is then obtained for systems in which all the three components are isomorphic with complete solid solubility at low temperature and complete liquid solubility at high temperature. Such a phase diagram is shown in Figure 6.6a. Here, the liquidus and solidus lines of the three binaries lie on the faces of the equilateral prism. However, few, if any, ternary phase diagrams are this simple.

A ternary system that is a little more complicated, yet relatively simple, in which all the three binary systems are eutectic with small solid solubility is shown in Figure 6.6b.

It is very common to represent the ternary phase diagrams in two dimensions. The conventional two-dimensional phase diagrams of ternaries are either complete projections of the three-dimensional diagrams down the temperature axis or projections of isothermal cuts. The resulting liquidus and solidus are then projected onto the composition base. This would be an equilateral triangle in which each corner represents 100% of the respective component, the opposite side corresponding to 0% (Figure 6.7a). All points on the lines parallel to a side represent compositions with constant concentration of the component identified with the opposite corner. The points on any line drawn through the corner to the opposite side will have a fixed ratio of

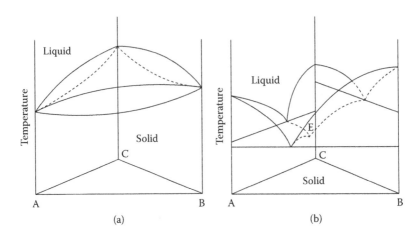

FIGURE 6.6
(a) Isomorphous ternary system and (b) ternary system with eutectics.

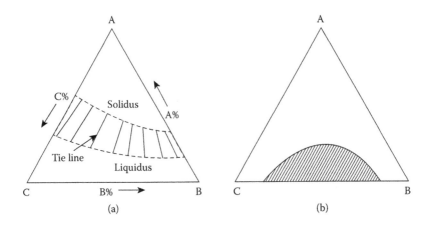

FIGURE 6.7
(a) Composition plane of a ternary system on which liquidus and solidus lines are projected and (b) a ternary system showing a miscibility gap between B and C.

the components identified with the other two corners, this ratio being equal to the one given by the point of intersection of that line with the opposite side.

An important difference between the isobaric binary and isobaric ternary systems is that the solidus and liquidus compositions are not uniquely specified by the temperature in the latter case. Also, the orientation of the tie lines is not defined, and one has to determine them through experiments only. In the absence of any experimental data one can only predict them as it has been done in Figure 6.7a.

Addition of a third component to a binary often alters the miscibility gap significantly. This is schematically shown in Figure 6.7b, which shows

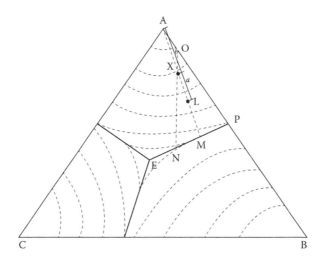

FIGURE 6.8
Two-dimensional representation of a ternary system with eutectics.

a miscibility gap between B and C in a ternary system ABC. The gap is pretty wide, but, as can be seen from Figure 6.7b, with the addition of the component A in BC they become more and more miscible.

Figure 6.8 shows a ternary system with eutectics in two dimensions as seen vertically from above in which the phase boundary curves are shown as full lines and isotherms as broken lines. Since E is a ternary eutectic point, the temperature decreases toward the center of the diagram. Let us now trace the crystallization history of a composition, say X shown in Figure 6.8, as we cool the system. We should note that the final solid must consist of crystals A + B + C with the initial composition X. At a temperature at which the liquid of composition X just intersects the liquidus surface, crystals of A begin to precipitate. As the temperature is lowered further, more and more crystals of A would be formed, and the composition of the liquid would move along a straight line away from A, thereby getting enriched in the components B + C. If we want to determine the relative proportion of crystals and liquid at any point on the line, say L, we can do so applying the lever rule.

Percentage of crystals = $[a/(a + b)] \times 100$
Percentage of liquid = $[b/(a + b)] \times 100$

On cooling further, the path of liquid composition will intersect the phase boundary curve at M at which point crystals of B will start forming. The liquid path now will be along the boundary curve toward N. During this period, a mixture of A + B will be crystallized and its composition will be as represented by point P. At N, the composition of the solid phases crystallized to that point can be found by drawing a straight line from N through X and extending it until it intersects the composition axis at O, which gives the proportion via the lever rule. That is

Percentage of the solid = (distance NX/distance ON) × 100

Percentage of the liquid = (distance OX/distance ON) × 100

We can determine the composition of all phases present in the system at this point. The liquid composition is the composition of point N as read from the basal triangle. Since it is a mixture of A, B, and C, it will be expressed in terms of the percentages of A, B, and C. The compositions of the solids A and B are 100% as they are pure solid phases.

On further cooling, the liquid composition will move toward the ternary eutectic, E, at which point crystals of C will start precipitating. The temperature will remain constant until the remaining liquid is solidified. The final crystalline product will consist of crystals of A + B + C in the same proportions as the initial composition X. The above illustration is rather simple and straightforward. The crystallization process in ternary systems that contain a compound that melts congruently or a compound that melts incongruently is much more complex.

Concentration–Temperature Phase Diagrams (Solubility Curves)

Crystal growth from aqueous solutions is probably the oldest known growth technique and is very widely used to produce a number of technologically important crystals. Since the growth is usually carried out at constant pressure (atmospheric pressure), the corresponding phase diagrams are in the concentration–temperature plane and are commonly called solubility curves. Solubility curves may have positive or negative slopes or sometimes practically zero slope, as shown schematically in Figure 6.9. The classic examples for these are sodium nitrate, sodium sulfate, and sodium chloride [9]. The difference in temperature dependence of solubility can be understood from the well-known Le Chatelier principle, which states that "if a change takes place in one of the thermodynamic variables (such as temperature, pressure, concentration) under which equilibrium exists, the system will tend to adjust itself so as to minimize the effect of that change" [3]. For example, changing the concentration of an ingredient will shift the equilibrium to the side that would reduce that change in concentration. This means, if the heat of solution of a solute in a particular solvent is endothermic (i.e., the heat is absorbed when the solution is formed), then one expects the solubility of that solute to increase with temperature. Taking the example of sugar solution, the heat of solution here is endothermic (that is why when sugar is added to water, the resulting solution is cooler than the solvent water); hence, the solubility of sugar in water increases with temperature. However, if the dissolution reaction is exothermic, then the solubility will decrease with temperature for that solute. Sodium sulfate–water system belongs to this category.

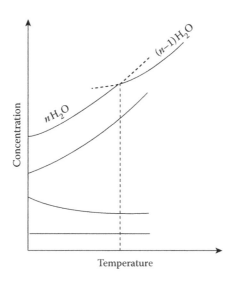

FIGURE 6.9
Temperature dependence of solubility.

It must be noted that several salts that are soluble in water exist with different water of crystallization and the content of water in the crystal depends on the temperature at which the crystal is grown. Because of this, only a limited number of salts can be grown in the anhydrous form. Usually, the higher the growth temperature, the lower the water content in the crystal. A salt with two different water contents, for example, may have a solubility curve with discontinuity, as also schematically shown in Figure 6.9. A typical example for such a system is nickel sulfate, which grows with seven molecules of water below 30.7°C and six molecules of water above this temperature [9].

In general, knowledge of solubility curves is quite crucial in deciding the method of achieving supersaturation and subsequent crystal growth. In many cases, the dissolution process is described by the van't Hoff equation

$$d\ln C/dT = \Delta H/RT^2 \qquad (6.9)$$

where C is the equilibrium concentration of the solute (solubility), T is the temperature in absolute scale, ΔH is the heat of solution, and R is the gas constant. When only limited solubility data are available, the van't Hoff equation can be used to produce or extend the solubility curves.

Slope of the solubility curve decides the conditions to be imposed during growth. For example, if the slope is positive and steep, the temperature has to be controlled with high precision while lowering because minor temperature fluctuation can lead to spontaneous nucleation. Materials with moderate slopes are therefore easier to grow. If the slope is negative, the temperature-raising method may have to be employed to achieve supersaturation required for growth. If the solubility curve is practically flat, one has to resort to solvent evaporation at constant temperature to grow the crystal.

Typical Phase Diagrams

In this section, we will discuss phase diagrams of a few representative crystals and see how they influence their growth processes. Since growth of crystals from the liquid phase is most widely practiced commercially and plays a very prominent role in the development of crystals for technological applications, the focus of this section will be on growth from aqueous solutions, from melts as also from high-temperature solutions.

Sodium *p*-Nitrophenolate

We discuss here the case of sodium *p*-nitrophenolate (NPNa), a semiorganic nonlinear optical crystal, to illustrate the importance of solubility curves in crystal growth. Sodium *p*-nitrophenolate, although soluble in water, was initially grown from methanol solution and the resulting crystals contained two molecules of water per formula unit $(NaO(C_6H_4NO_2) \cdot 2H_2O)$ as water of crystallization [10,11]. The same material when grown from water at room temperature (30°C) yielded transparent crystals with different morphology. However, these crystals lost their transparency on exposure to the atmosphere (Figure 6.10). From thermal and structural studies, this crystal was identified as sodium *p*-nitrophenolate tetrahydrate $(NaO(C_6H_4NO_2) \cdot 4H_2O)$, which crystallizes in space group $P2_1/c$ [12]. Since the content of water (water of crystallization) depends on the growth temperature, attempt was made to grow the crystal from aqueous solution at various temperatures above 30°C. Various test runs were carried out in the temperature range of 35°C–55°C. It was observed that growth at temperatures between 55°C and 38°C yielded only the dihydrate form of the crystal. Test runs at 37°C, on the contrary, gave both forms, that is, dihydrate as well as tetrahydrate. At 36°C, all the crystals obtained were tetrahydrate. Hence, from the detailed test runs carried out, it was possible to determine the temperature (38°C) above which one can surely obtain only the dihydrate form of NPNa. Once the range of temperature in which the two forms would grow was established, detailed solubility characteristics of the tetrahydrate and dihydrate forms were determined gravimetrically in their respective temperature regimes. The solubility of NPNa·2H₂O in water was measured as a function of temperature in the temperature range of 40°C–75°C and that of NPNa·4H₂O in the range of 25°C–35°C. Figure 6.11 shows the solubility curves of both NPNa·2H₂O and NPNa·4H₂O. In Figure 6.11, the region in which both forms

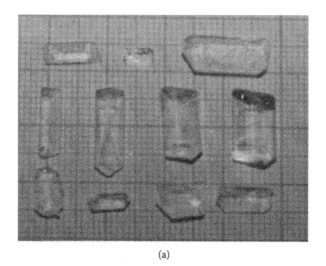

(a)

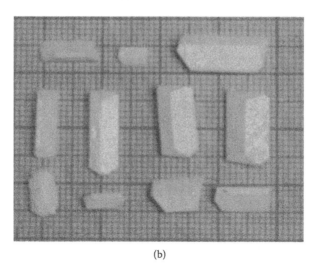

(b)

FIGURE 6.10
Spontaneously grown crystals of NPNa obtained from aqueous solution at 30°C (a) freshly removed from solution and (b) after 3 h of exposure to the atmosphere. (Reprinted from *Journal of Crystal Growth*, 275, Vanishri, S., et al., Crystal growth of semiorganic sodium *p*-nitrophenolate dihydrate from aqueous solution and their characterization, E141–E146, 2005, with permission from Elsevier.)

would simultaneously grow is shown by hatched lines. Once the temperature region in which the dihydrate form grows is determined, large crystals of NPNa·2H$_2$O could be grown from aqueous solution by the temperature-lowering method [13]. Figure 6.12a shows the photograph of the single crystals thus grown from aqueous solution. The crystals grown from methanol solution are also shown here for comparison (Figure 6.12b).

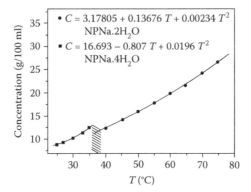

FIGURE 6.11
Solubility plots of NPNa dihydrate and tetrahydrate in water. (Reprinted from *Journal of Crystal Growth*, 275, Vanishri, S., et al., Crystal growth of semiorganic sodium *p*-nitrophenolate dihydrate from aqueous solution and their characterization, E141–146, 2005, with permission from Elsevier.)

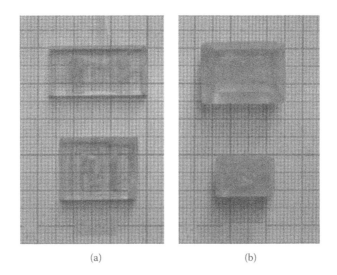

(a)   (b)

FIGURE 6.12
Crystals of NPNa-$2H_2O$ grown from (a) water and (b) methanol. (Reprinted from *Journal of Crystal Growth*, 275, Vanishri, S., et al., Crystal growth of semiorganic sodium *p*-nitrophenolate dihydrate from aqueous solution and their characterization, E141–E146, 2005, with permission from Elsevier.)

Gallium Antimonide

III-V ternaries and quaternaries lattice-matched to gallium antimonide (GaSb) have turned out to be promising candidates for high-speed electronic and long-wavelength photonic devices. Consequently, there has been tremendous thrust in research and developmental activities in GaSb-based systems, including growth of large and high-quality GaSb crystals [14].

The phase diagram of GaSb has been determined independently by Koster and Thoma [15] and Greenfield and Smith [16]. Later on, the liquidus has been reevaluated in different regions of the phase diagram by several researchers. The phase diagrams of GaSb are shown in Figure 6.13 [14,17]. The melting point of GaSb has been reported to lie between 705°C and 712°C. A melting point depression of less than 50°C is observed for compositions ±30 atomic% on either side of the stoichiometric composition. The dissociation pressure

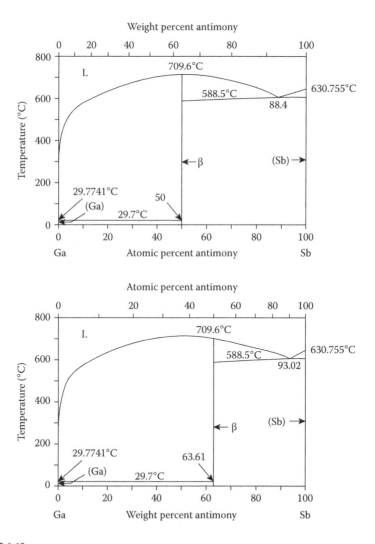

FIGURE 6.13
Phase diagrams of GaSb. (Reprinted with permission from Dutta, P. S., et al., The physics and technology of gallium antimonide: an emerging optoelectronic material, *Journal of Applied Physics*, 81, 5821–5870, 1997, American Institute of Physics; Hansen, M., ed., *Constitution of binary alloys*, McGraw Hill, New York, 1958, p. 755. With permission.)

at the melting point of GaSb is about 10^{-2} mm Hg. At above 370°C, Sb starts volatilizing from the melt. At the maximum melting point, the partial vapor pressure of Sb is ~3×10^{-6} Torr [18]. With this partial pressure, up to ~2×10^{15} Sb atoms per second could be lost from each square centimeter of solid surface. The partial pressure of Ga is less than 10^{-9} Torr at the maximum melting point [18]. Thus, in a 10 h growth run, ≈10^{-3} mol of Sb would be lost, which amounts to ~0.1% change in the Sb/Ga ratio. Hence, usually during synthesis of GaSb, the Sb/Ga ratio in the melt is 1.001. The binary compounds GaSb and InSb are totally miscible, both in liquid and in solid state, thereby allowing for the physical properties of the InGaSb alloy to vary continuously with the InSb composition [19]. The equilibrium phase diagram in the GaSb-InSb quasi-binary section of the GaInSb system is shown in Figure 6.14 [20]. During the solidification of InGaSb, there is a rejection of InSb into the liquid. The rejected material accumulates in front of the interface and spreads into the liquid phase, by diffusion and mixing induced by convection. Consequently, polycrystalline ingots obtained by directional solidification are with grain structure continuously changing along the growth direction due to variation in crystal composition by segregation [21]. One of the technical difficulties encountered by early workers is the long time required for reaching equilibrium of this compound when prepared from powdered precursors. Extremely long mixing times of the order of even several hundreds of hours was found to be insufficient to prepare this compound in homogeneous form.

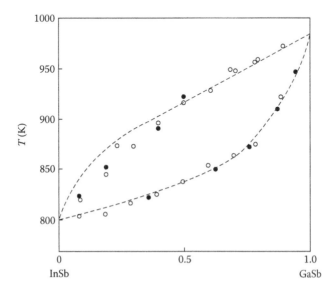

FIGURE 6.14

Pseudo-binary phase diagram of InSb-GaSb. (Reprinted from *Journal of Crystal Growth*, 32, Ansara, I., et al., Étu de thermodynamique du systèmeternaire gallium-indium-antimoine, 101–110, 1976, with permission from Elsevier.)

Lithium Niobate

Lithium niobate ($LiNbO_3$) is one of the most thoroughly studied ferroelectrics. It was first grown and recognized to be ferroelectric by Matthias and Remeika in 1949 [22]. It was rediscovered for its unique electro-optic properties in 1964 by Peterson et al. [23]. In the same year, Boyd et al. [24] showed that $LiNbO_3$ has a high nonlinear optical coefficient and adequate birefringence to phase match for second harmonic and parametric generation. Due to its high electro-mechanical coupling coefficient and low loss at ultrasonic frequencies, $LiNbO_3$ has also been increasingly used for acoustic applications [25].

The lithium niobate phase exhibits a solid solubility ranging from compositions near stoichiometric value (50 mol%) to lithium-poor compositions as low as approximately 45 mol% Li_2O (Figure 6.15a) [26–29]. The composition corresponding to 48.45 mol% of Li in Figure 6.15b is the congruent composition [30]. The dashed line in the phase diagram shows the dependence of ferroelectric transition temperature on crystal composition. Crystals grown with congruent composition have exactly the same cationic ratio as the melt, and the liquid composition remains unaltered during the growth process. Consequently, this has been the preferred composition for most research and applications. Crystal growth from any other liquid composition leads to compositional inhomogeneity along the pulling direction.

Since the stoichiometric composition does not melt congruently, lithium niobate crystals with exact stoichiometry are difficult to produce. Several methods have been explored to grow stoichiometric lithium niobate crystals, the most suitable one being the double-crucible Czochralski method with sophisticated automatic powder feeding technique [31]. This method exploits the coexistence of near stoichiometric $LiNbO_3$ with Li-rich melt in the phase diagram (the dashed horizontal line in Figure 6.15b).

A much simpler method similar to top seeded solution growth technique is the growth from conventional a Czochralski system with excess Li_2O (58.6 mol%). Since it is a self-flux method, there is no contamination due to the flux. The disadvantage of this method is that the growth yield is not more than 15% of the total charge. This constraint arises due to continuously shifting melt composition toward the eutectic during the growth.

Cesium Lithium Borate

Cesium lithium borate (CLBO) was developed in the mid-1990s to facilitate the generation of coherent radiation in the ultraviolet region through nonlinear optical processes [32]. Although excimer lasers produce significant power in the deep ultraviolet region, these lasers are bulky, involve corrosive gases, and require regular maintenance. (The term *excimer* is short form for "excited dimer." Most "excimer" lasers are of the noble gas halide type.) A maintenance-free, compact, all-solid-state laser was always preferred, and CLBO was developed with this background.

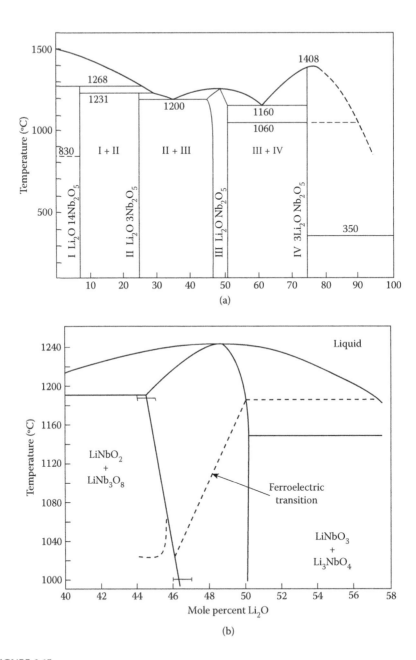

FIGURE 6.15
(a) Complete phase diagram of LiNbO₃ and (b) schematic phase diagram of pseudo-binary phase diagram of the Li₂O-Nb₂O₅ system near the congruent and stoichiometric composition of LiNbO₃. (Panel a is reprinted from *Journal of Crystal Growth*, 18, Svaasand, L. O., et al., Crystal growth and properties of LiNb₃O₈. 179–184, 1973, with permission from Elsevier; Panel b is from O'Bryan, H. M., et al., *Journal of the American Ceramic Society*, 68, 493–496, 1985.)

The pseudo-binary phase diagrams of CLBO as determined by Sasaki et al. [33] are shown in Figure 6.16. From these, it is clear that CLBO melts congruently at stoichiometric composition at 848°C; hence, in principle melt method can be employed for crystal growth. However, viscosity of the stoichiometric melt is very high, and it hinders the atomic diffusion (material transport) during crystal growth. Examination of the phase diagram also shows that CLBO crystals can be grown from fluxes that are either rich or poor in B_2O_3. However, growth from B_2O_3-rich flux is even more difficult due to its increased viscosity. To overcome this problem, CLBO has been grown with boron-deficient charge of $Cs_2O{:}Li_2O{:}B_2O_3 :: 1{:}1{:}5.5$, as this composition has lower viscosity compared to stoichiometric melt.

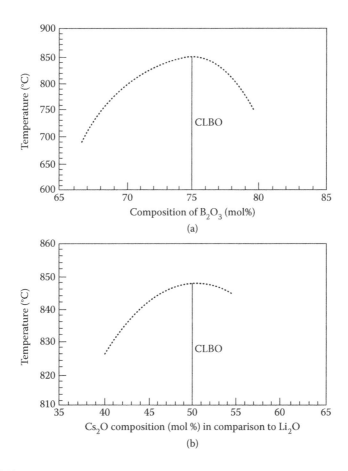

FIGURE 6.16

Pseudo-binary phase diagrams of the (a) $(Cs_2O{:}Li_2O)$-B_2O_3 system and (b) Cs_2O-Li_2O system at 75% B_2O_3 in the ternary system $Cs_2O{:}Li_2O{:} 3 B_2O_3$. (Reprinted from *Materials Science and Engineering*, R30, Sasaki, T., et al., Recent development of nonlinear optical borate crystals: key materials for generation of visible and UV light, 1–54, 2000, with permission from Elsevier.)

Potassium Titanyl Phosphate (K₂O–P₂O₅–TiO₂ System)

As an example of a ternary phase diagram, we discuss the phase diagram of potassium titanyl phosphate (KTP) with the chemical formula $KTiOPO_4$. KTP is a unique nonlinear optical crystal with exceptional properties [34]. It combines large nonlinearity, large electro-optic coefficients, and high laser damage threshold, and it enables good phase matching and possesses high optical homogeneity. In conjunction with its isomorphs, it offers exceptional versatility for device fabrication.

KTP is grown by melt or high-temperature solution growth technique. The high-temperature phase diagram of KTP was first studied by Voronkova and Yanovskii [35] and later in detail by Iliev et al. [36]. They determined the concentration regions and crystallization temperature of the $KTiOPO_4$ phase. Figure 6.17 is a portion of the ternary phase diagram that shows a plot of concentration region and spontaneous crystallization temperature of KTP in $K_2O–P_2O_5–TiO_2$ solvent system for different compositions. Figure 6.17 shows the region of stable KTP phase in this ternary system. It is also seen that KTP phase has a reasonably wide crystallization region. This region is surrounded by regions where KPO_3, $K_4P_2O_7$, $K_2Ti_6O_{13}$, TiO_2, and $KTi_2(PO_4)$ are the sole crystallizing phases. The phases KPO_3 and $K_4P_2O_7$ crystallize from solutions with very low TiO_2 content.

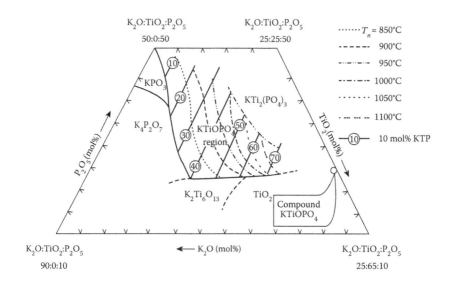

FIGURE 6.17
Phase diagram of $K_2O–P_2O_5–TiO_2$ system showing the crystallization regions of KTP. (Reprinted from *Journal of Crystal Growth*, 100, Iliev, K., et al., Physicochemical properties of high-temperature solutions of the K_2O-P_2O_5-TiO_2 system suitable for the growth of $KTiOPO_4$ (KTP) single crystals, 219, 1990, with permission from Elsevier.)

It should be emphasized that the presence of dopant materials intended to be incorporated in the crystal often significantly alters the stability region in the phase diagram. For example, the crystallization regions of rare earth–doped KTP (RE: Nd, Th, Ho, Er, Tm, and Yb) are significantly narrower than those of undoped KTP [37]. The stable KTP phase appears only below a critical RE concentration. It has also been found that the highest concentration of RE in the crystal strongly depends on the RE ionic radius.

Summing up, it is clear that a knowledge and understanding of phase diagrams are crucial for both the selection of growth technique as well as imposing and adjusting several growth-related parameters during growth. It is likely that not all the published phase diagrams are accurate. Nevertheless, they contain useful information for a crystal grower. Our discussion on practical phase diagrams was limited to only a few simple systems. One may have to encounter more complicated systems and at times may have to determine the phase diagram first before proceeding for growth. Phase diagrams are usually determined by a combination of experimental techniques such as differential thermal analysis, differential scanning calorimetry, thermo-gravimetric analysis, elemental analysis, and x-ray diffraction for phase analysis. Such facilities should be available in and around a crystal growth laboratory.

References

1. Gordon, P. 1968. *Principles of phase diagrams*. New York, NY: McGraw Hill.
2. Rosenberger, F. 1979. *Fundamentals of crystal growth I*. Berlin, Germany: Springer.
3. Rao, C. N. R., and K. J. Rao. 1978. *Phase transitions in solids*. New York, NY: McGraw Hill.
4. Muhlberg, M. 2008. Phase diagrams for crystal growth. In *Crystal growth technology*, edited by H. J. Scheel and P. Capper. Weinheim, Germany: Wiley-VCH Verlag GmbH, pp. 3–26.
5. Harrison, D. E., and W. A. Tiller. 1965. Growth of large single crystals of hexagonal selenium from the melt at high pressures. *Journal of Applied Physics* 36: 1680–1683.
6. Gibbs, J. W. 1961. *Scientific papers*. New York, NY: Dover.
7. Bundy, F. 1979. The P, T phase and reaction diagram for elemental carbon. *Journal of Geophysical Research* 85(B12): 6930–6936.
8. Steinbeck, J., G. Braunstein, M. S. Dresselhaus, T. Venkatesan, and D. C. Jacobson. 1985. A model for pulsed laser melting of graphite. *Journal of Applied Physics* 58: 4374–4382.
9. Holden, A., and P. Singer. 1961. *Crystals and crystal growing*. London, UK: Heinemann.
10. Minemoto, H., Y. Ozaki, N. Sonoda, and T. Sasaki. 1994. Crystal growth and the nonlinear optical properties of 4-nitrophenol sodium salt dihydrate and its deuterated material. *Journal of Applied Physics* 76: 3975–3980.

11. Brahadeeswaran, S., V. Venkataramanan, J. N. Sherwood, and H. L. Bhat. 1998. Crystal growth and characterization of nonlinear optical material: Sodium p-nitrophenolate dihydrate. *Journal of Materials Chemistry* 8: 613–618.
12. Brahadeeswaran, S., H. L. Bhat, J. N. Sherwood, and R. M. Vrcelj. 2002. Sodium p-nitrophenolate tetrahydrate. *Acta Crystallographica* E58: m290–m292.
13. Vanishri, S., S. Brahadeeswaran, and H. L. Bhat. 2005. Crystal growth of semi-organic sodium *p*-nitrophenolate dihydrate from aqueous solution and their characterization. *Journal of Crystal Growth* 275: E141–E146.
14. Dutta, P. S., H. L. Bhat, and V. Kumar. 1997. The physics and technology of gallium antimonide: An emerging optoelectronic material. *Journal of Applied Physics* 81(9): 5821–5870.
15. Koster, W., and B. Thoma. 1955. Constitution of the gallium-antimony, gallium-arsenic and aluminum-arsenic systems. *Zeitschrift für Metallkunde* 46: 291–293.
16. Greenfield, I. G., and R. L. Smith. 1955. Gallium-antimony system. *Transactions of the Metallurgical Society of AIME* 203: 351–353.
17. Hansen, M., ed. 1958. *Constitution of binary alloys.* New York, NY: McGraw Hill, p. 755.
18. Sunder, W. A., R. L. Barns, T. Y. Kometani, J. M. Parsey, Jr., and R. A. Laudise. 1986. Czochralski growth and characterization of GaSb. *Journal of Crystal Growth* 78: 9–18.
19. Wooley, J. C., B. A. Smith, and D. G. Lees. 1956. Solid solution in GaSb-InSb system. *Proceedings of the Physical Society of London* B69: 1339–1343.
20. Ansara, I., M. Gambino, and J. P. Bros. 1976. Étu de thermodynamique du systèmeternaire gallium-indium-antimoine. *Journal of Crystal Growth* 32: 101–110.
21. Joullie, A., A. Allegre, and G. Bougnot. 1972. Preparation and electrical properties of homogeneous $Ga_xIn_{1-x}Sb$ alloys. *Materials Research Bulletin* 7: 1101–1107.
22. Matthias, B. T., and J. P. Remeika. 1949. Ferroelectricity in the ilmenite structure. *Physics Review* 76: 1886–1887.
23. Peterson, G. E., A. A. Ballman, P. V. Lenzo, and P. M. Bridenbaugh. 1964. Electro-optic properties of $LiNbO_3$. *Applied Physics Letters* 5: 62–63.
24. Boyd, G. D., R. C. Miller, K. Nassau, W. L. Boyd, and A. Savage. 1964. $LiNbO_3$: An efficient phase matchable nonlinear optical material. *Applied Physics Letters* 5: 234–235.
25. Spencer, G. E., P. V. Lenzo, and A. A. Ballman. 1967. Dielectric materials for electrooptic, elastooptic, and ultrasonic device applications. *Proceedings of the IEEE* 5: 2074–2108.
26. Lerner, P., C. Legras, and J. P. Dumas. 1968. Stoechiométrie des monocristaux de métaniobate de lithium. *Journal of Crystal Growth* 3/4: 231–235.
27. Carruthers, J. R., G. E. Peterson, M. Grasso, and P. M. Bridenbaugh. 1971. Nonstoichiometry and crystal growth of Lithium Niobate. *Journal Applied Physics* 42: 1846–1851.
28. Svaasand, L. O., M. Eriksrud, A. P. Grande, and F. Mo. 1973. Crystal growth and properties of $LiNb_3O_8$. *Journal of Crystal Growth* 18: 179–184.
29. Svaasand, L. O., M. Eriksrud, G. Nakken, and A. P. Grande. 1974. Solid-solution range of $LiNbO_3$. *Journal of Crystal Growth* 22: 230–232.
30. O'Bryan, H. M., P. K. Gallagher, and C. D. Brandle. 1985. Congruent composition and Li-Rich phase boundary of $LiNbO_3$. *Journal of the American Ceramics Society* 68: 493–496.

31. Kitamura, K., J. K. Yamamoto, N. Iyi, S. Kimura, and T. Hayashi. 1992. Stoichiometric LiNbO$_3$ single crystal growth by double crucible Czochralski method using automatic powder supply system. *Journal of Crystal Growth* 116: 327–332.
32. Sasaki, T., Y. Mori, I. Kuroda, S. Nakajima, K. Yamaguchi, S. Watanabe, and S. Nakai. 1995. Caesium lithium borate: A new nonlinear optical crystal. *Acta Crystallographica Section* C 51: 2222–2224.
33. Sasaki, T., Y. Mori, M. Yoshimura, Y. K. Yap, and T. Kamamura. 2000. Recent development of nonlinear optical borate crystals: Key materials for generation of visible and UV light. *Materials Science and Engineering* 30: 1–54.
34. Satyanarayana, M. N., A. Deepthy, and H. L. Bhat. 1999. Potassium titanyl phosphate and its isomorphs: Growth, properties and applications. *Critical Reviews in Solid State Materials Sciences* 24(2): 103–189.
35. Voronkova, V. I., and V. K. Yanovskii. 1988. Flux growth and properties of the KTiOPO$_4$ family crystals. *Inorganic Materials* 24(2): 273–277.
36. Iliev, K., P. Peshev, V. Nikolov, and I. Koseva. 1990. Physicochemical properties of high-temperature solutions of the K$_2$O-P$_2$O$_5$-TiO$_2$ system suitable for the growth of KTiOPO$_4$ (KTP) single crystals. *Journal of Crystal Growth* 100: 219–224.
37. Sole, R., V. Nikolov, I. Koseva, P. Peshev, X. Ruiz, C. Zaldo, M. J. Martin, M. Augilo, and F. Diaz. 1997. Conditions and possibilities for rare-earth doping of KTiOPO$_4$ flux-crown single-crystals. *Chemistry of Materials* 9: 2745–2749.

Section II

Practice

7

Growth from the Solid Phase

Introduction

In Section I of this book, we dealt briefly with various theories and mechanisms of crystal growth. However, an experimentalist has always felt that the rate of growth and perfection of the grown crystal are more influenced by careful and judicious control of the growth parameters than by the detailed knowledge of mechanisms and theories related to the growth. Nevertheless, they are extremely important for the development of the subject and also have applications, since it has been evidenced in recent years that the growth mechanisms do influence certain crystal properties.

A large number of growth techniques have been developed over the years to grow a wide variety of crystals. Each of these techniques has its own merit and applicability. In an introductory book like this, it is rather impossible to discuss in detail of these techniques, as the literature pertaining to each of them is quite voluminous. However, we deal with the basics and the procedures involved in some of the important growth methods.

In the literature, one finds that the growth techniques have been classified in a number of ways, and there is no unique way of classifying them. For example, the existing growth techniques may be broadly classified under the following four headings:

1. Growth from the solid phase
2. Growth from the melts
3. Growth from the liquid solution
4. Growth from the vapor phase

In this book, we follow this classification.

Classification based on the technique in producing bulk crystals or thin films is often found in the literature. In this book, thin film growth is discussed separately in Chapter 11. The growth techniques have also been categorized as single (mono)-component and multi (poly)-component techniques depending on the number of components present in the growing medium. For example, melt growth techniques are essentially single-component techniques in the sense that here the material to be grown as a single crystal itself is the precursor. However, solid-state precipitation or exsolution is an example of polycomponent solid–solid growth. Growth from a liquid solution invariably has more than one component; hence, these are polycomponent growth methods. Crystal growth from a vapor phase may involve either monocomponent or polycomponent techniques. Strictly speaking, the monocomponent technique should not include any systems where a second component, either accidental or intentional, is present. However, a practical definition of monocomponent growth would allow the presence of impurities and dopants, provided their concentration is low enough so that diffusion processes are not of significance. On the contrary, if a second component is deliberately added to lower the melting point of the material to be crystallized, the growth should be considered as polycomponent.

Solid–Solid Growth Methods

The greatest advantage of the solid–solid growth methods lies in their simplicity. In addition, there is less possibility of contamination, the shape of the grown crystal can be predetermined, and distribution of impurities can be fixed before growth. All of this is because the starting material is a solid and what is being done in principle is that the sample is simply annealed. Of course, this is an oversimplified statement, as we will realize soon. For an extensive coverage of the topic, the reader is advised to refer to the book by Humphreys and Hatherly [1]. The main disadvantage, however, is that it is difficult to control nucleation in the solid phase and hence difficult to obtain large single crystals. Also, achieving a high degree of structural perfection is difficult. Even under ideal growth conditions, some crystallographic defects will always be present in the grown crystals. Lack of reproducibility is another problem with these techniques. Nevertheless, solid–solid growth methods are useful to grow single crystals or, more often, materials that are predominantly single crystalline and hence are being practiced.

Of the many techniques available under this category, grain growth, strain-annealing, and solid-state transformation techniques are quite popular, and these are discussed in the sections that follow.

Grain Growth

In grain growth technique, the size of crystalline grain increases when a finely grained material is kept at high temperature for certain duration. This is beautifully illustrated in a classic photograph of two lead ingots (Figure 7.1), one as cast and the other as cast and annealed for 1 h at 300°C [2]. This grain

(a)

(b)

FIGURE 7.1
Zone-refined lead ingots (a) as cast and (b) as cast and annealed at 300 °C for 1 h. (From Aust, K. T., *Art and science of growing crystals*, John Wiley & Sons, New York, London, 1963, pp. 452–478. With permission.)

growth occurs after the completion of recovery and recrystallization processes that bring down the internal energy of the system. (Recrystallization is a process by which deformed grains are entirely replaced by a new set of undeformed grains that nucleate and grow in the matrix.) Any further reduction in the internal energy would be possible only by reducing the total area of the grain boundary.

It is well established that amorphous solids have excess free energy with respect to their crystalline phases, and this excess free energy is the driving force in the case of crystal growth from amorphous solids. Likewise, polycrystalline materials also have higher free energy with respect to their single crystalline counterparts. As we have seen, the boundary between two grains is a crystalline defect, and so it is associated with a certain amount of energy. As a result, there is a thermodynamic driving force for the reduction in total area of the boundary. As the grain size increases at the expense of surrounding smaller grains, there will be reduction in the actual number of grains. Consequently, the total area of boundary will decrease. However, since energy available to drive grain growth is very low compared to that available for, say, liquid to solid transformation, grain growth normally occurs at a much lower rate, and hence it takes much longer time to attain appreciable sizes.

Grain growth is usually studied by the examination of sectioned, polished, and etched samples under optical microscope. Through such microscopic examination, a large amount of data were collected from which certain general inferences have been drawn. First, it has been established that grain growth occurs by the movement of grain boundaries and not by coalescence of grains. It is observed that such movements may be discontinuous, leading to discontinuous grain growth. During grain growth, the direction of motion of the boundary may change suddenly. The growth of a particular grain occurs at the expense of smaller grains around it, and the rate of consumption of smaller grains often increases when these are about to be completely consumed. If a grain boundary is curved, its migration is preferably toward its center of curvature. Also, grain boundary energy is orientation dependent. Thus, the driving force for growth of a particular grain is due to both the size difference and the orientation difference. Further, presence of pronounced texture in a material indicates that most of the crystallites are preferentially oriented. Another point to be considered is that the initial grain size must be sufficiently small and uniform so that once a nucleus grows larger than its immediate neighborhood it would be able to consume the rest.

Discontinuous grain growth is characterized by a subset of grains growing at a high rate at the expense of their neighbors, resulting in a texture dominated by a few very large grains. This is sometimes referred to as exaggerated grain growth in which the growth of most of the grains in the matrix is inhibited. For this to happen, the subset of grains must possess some advantage over their competitors such as a high grain boundary energy, locally high grain boundary mobility, favorable texture, or lower local second-phase particle density. The presence of a small amount of liquid seems to promote exaggerated grain growth. For example, due to the local presence of a small amount of liquid at 1000 °C, a few grains grew to abnormally large size in YBCO superconducting ceramics [3]. By placing a small amount of SiO_2 on the top of a compacted $BaTiO_3$ ceramic specimen, crystals as large as 1.5 cm were produced by Yoo et al. [4].

In the case of grain growth where boundary motion is driven only by the reduction of the total amount of grain boundary surface energy (i.e., additional contributions to the driving force such as elastic strains, temperature gradients, etc. are neglected), one can show that the linear boundary velocity approximately follows an exponential law [5] of the type,

$$v = v_0 \, exp(-Q/RT) \qquad (7.1)$$

where v_0 is a proportionality constant, T is the absolute temperature, and Q is the activation energy for boundary mobility. For this reason, grain growth is appreciable only at temperatures $\sim(0.5\text{--}0.7) \, T_m$, with T_m being the melting point.

In general, this equation is found to hold only for high-purity materials but rapidly fails when even a small amount of impurity is introduced.

In solid–solid growth, in many cases, grain growth might occur without the need of nucleation because the already present grains can act as nuclei. In fact, one of the main problems in this method is to control the number of growth centers because in a polycrystalline specimen many grains can become potential growth centers. To overcome this problem, many methods have been devised. For example, seeding can be done by welding a large grain onto a polycrystalline ingot and causing the welded grain to grow into the ingot at the expense of smaller grains present in it. Annealing the poly-crystalline sample in a temperature gradient is another method to control the number of growth centers. Because crystal growth occurs more readily than nucleation, an imposed temperature gradient favors the continued growth of an existing grain rather than formation of fresh nuclei. Cyclic heat-ing method has also been used to suppress the undesired grains to nucleate and grow. Even though grain growth may occur with the existing grains in a polycrystalline sample, under some conditions it has been seen that nucle-ation of new grains does occur during annealing, and when that happens, it is these new grains that grow further at the expense of their neighbors.

Figure 7.2 shows the schematic of a grain growth setup originally due to Andrade [6] and particularly useful to grow single crystals in the form of wire or thin rods. Using such a setup, Andrade was able to grow single crys-tal wires of molybdenum and tungsten. In a typical growth run, an etched wire or a thin rod is held at its two ends by suitable fixtures. The specimen is usually kept in an inert atmosphere to prevent undesirable oxidation of the surface. Initially, the whole specimen is heated to the desired temperature

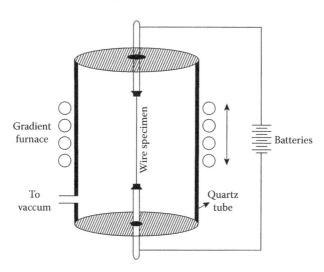

FIGURE 7.2
Schematic of a grain growth setup.

(i.e., below the crystallization temperature, which varies widely between 55% and 99% of the actual melting temperature). This can be done by simply passing current through the specimen. Subsequently, a hot zone producing the crystallization temperature is passed along the length of the specimen using a subsidiary ring heater. This heater can be either a resistive or an induction type. The rate of movement of the hot zone also varies widely and can be anywhere between 3 and 50 mm per hour.

In the literature, we see ample examples of preparation of crystals by the grain growth method, notably metals and alloys, although many of these results truly may be examples of growth by the strain-annealing technique that is discussed next. Inherently present strain in the sample or inadvertently introduced strain during sample preparation and processing may have aided the grain growth process in such cases.

Strain-Annealing Technique

Although the grain growth method has yielded good results, introduction of strain into the sample, leading to plastic deformation and then annealing it, significantly improves the results. This technique has been referred to in literature as the strain-annealing technique. A detailed account of the strain-annealing technique has been given by Laudise [7] in chapter 4 of his book *The Growth of Single Crystals*. Here, we give only the salient features of this method.

We saw in Chapter 3 that when a body is stressed beyond its elastic limit, withdrawal of the applied stress would not lead to the removal of complete strain and the body will get permanently deformed (i.e., plastically deformed). Plastically deformed bodies contain a large amount of strain energy. This stored energy is the main driving force for recrystallization in the strain-annealing technique.

In this method one generally starts with fine-grained matrix that is subjected to a minimal strain just sufficient to cause the formation of one or at most a few new grains during annealing. It can be seen from Figure 7.3 that the grain size increases with decreasing amount of prior strain down to a certain critical value below which no grain growth occurs [2]. In principle, a single crystal can be obtained by annealing a polycrystalline specimen that has been given a critical amount of strain. If less than critical strain is applied, no grain growth will occur during annealing. For larger strains, several crystallites will result in the same specimen. Critical strain varies from material to material and is in the range of 1%–10%. The critical strain is usually determined by straining a tapered specimen by applying suitable stress and thereby producing a strain gradient in the specimen. Annealing this specimen and subsequently examining it under the microscope after etching would reveal the region where the specimen has been strained critically (where the grains are larger and fewer). By measuring the area of cross-section of the specimen corresponding to that region, one can

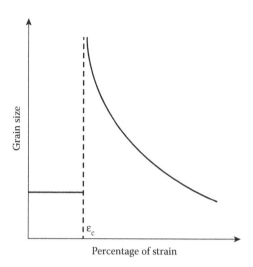

FIGURE 7.3
Grain size as a function of amount of prior strain. (From Aust, K. T., *Art and science of growing crystals*, John Wiley & Sons, New York, London, 1963, pp. 452–478. With permission.)

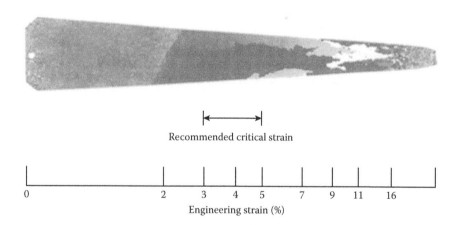

FIGURE 7.4
Portion of the tensile specimen of α-iron used to determine the critical strain for crystal growth. (With kind permission from Springer: *Metallurgical Transactions A*, Improved strain-anneal crystal growth technique, 6A, 1975, 403–408, Bailey, D. J., and Brewer, E. G.)

determine the critical strain. Figure 7.4 shows the portion of the tapered specimen of α-iron (subjected to tensile stress followed by annealing), used to determine the critical strain for crystal growth [8]. The region where larger and fewer grains are formed is indicated.

For crystal growth, one applies a temperature gradient (Figure 7.5). The specimen, which may have a tapered end to facilitate localized nucleation,

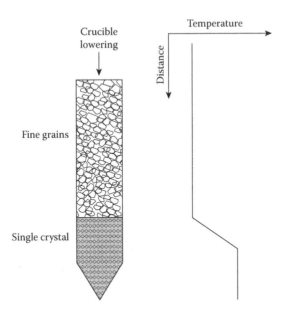

FIGURE 7.5
Lowering of the strained specimen with tapered end through a temperature gradient.

is made to pass through a furnace at a certain speed. The temperature gradient during growth is necessary to avoid simultaneous nucleation in different parts of the specimen. It is most favorable if, on entering the hot zone of the furnace, only one nucleus develops and grows fast enough to consume subsequent fine-grained material that enters the furnace before a second nucleus can be formed. The right gradient and velocity depend on the temperature dependence of the rates of nucleation and growth.

To promote localized nucleation at the tip of the specimen that enters the hot zone first, this part is heavily deformed by severely compressing or pinching it. This adds locally to overall critical deformation given to the specimen. When a severely deformed end enters a hot zone, several crystals form, which on further growth compete with one another until only one of them survives due to faster rate of growth. The temperature gradient ensures that additional nucleation does not occur ahead of the main growth front. The temperature, in effect, travels along the specimen, thereby increasing the probability that a single grain will travel along with it, consuming the deformed matrix before new grains nucleate. The temperature gradient should be as large as possible in order to decrease the probability of forming many new grains by recrystallization.

Despite many precautions taken prior to actual growth, the successful growth of a single crystal by this technique is often terminated by the random nucleation of several grains more favorably oriented for rapid grain growth. One of the useful ways of preventing the formation of parasitic nuclei ahead of the growth interface, which was employed successfully by Bailey and Brewer [8] to grow iron single crystals, is the use of the pulse heating method. With the pulse heating, the random nucleation during the growth of an intentionally introduced seed crystal could be suppressed because there is an induction period before the formation of a fresh critical nuclei, whereas there is no incubation period for the growth of an intentionally introduced seed. By properly adjusting the cyclic periods such that they are less than the induction period for nucleation, the seed can be propagated through the entire specimen. Cyclic heating would then lead to intermittent growth of the desired seed crystal. Bailey and Brewer applied pulse heating by alternately lowering and raising the work piece through the hot zone of the furnace set for a peak specimen temperature of 875°C. Although the specimen experienced a temperature gradient with this method of heating, the gradient was less than those used commonly by gradient annealing technique. Figure 7.6 [8] shows single crystals of α-iron in the form of sheet and rod grown by these authors.

Difficulties are generally encountered in attempts to grow crystals of high-purity materials by the strain-annealing method because, in the case of high-purity material, it is more difficult to attain a fine-grained stable matrix structure prior to introducing critical strain. This is primarily because grain growth can occur more readily in them. However, it has been found that by introducing a strong preferred orientation into the matrix one is able to stabilize the texture.

The strain-annealing method does not require any sophisticated setup. Furnaces capable of producing the required gradients and with various ramp-up and ramp-down rates are adequate. In addition, the furnace should be capable of holding different atmospheres. To strain the specimens, many methods are used. Strain may be introduced by compression, tension, torsion,

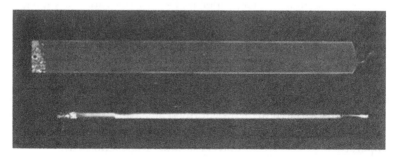

5.08 cm

2.0 in.

FIGURE 7.6

Single crystals of α-iron in the form of a sheet and rod grown by the strain-annealing technique using pulse heating. (With kind permission from Springer: *Metallurgical Transactions A*, Improved strain-anneal crystal growth technique, 6A, 1975, 403–408, Bailey, D. J., and Brewer, E. G.)

drawing, and extrusion. Tensile test apparatus, if available, would be very handy. Rolling can also be employed if the specimen is in the form of sheet.

The strain-annealing technique has been applied extensively to iron and its alloys. It has also been used to grow crystals of several metals like tantalum [9], niobium [10], and titanium [11].

Solid-State Phase Transformation

In growth by solid-state transformation, one starts with a poly- or single crystalline specimen of the undesired polymorph (usually a high-temperature phase) and transforms it to the required polymorph through a structural transition. To effect the desired phase change, one may have to apply pressure, vary the temperature, or both. A classic example of these two variables being simultaneously applied is the growth of diamond through polymorphic transformation of graphite.

In the case of materials undergoing phase transition, the grains of high-temperature phase are completely replaced by a new set of grains belonging to the low-temperature phase. If now the phase change takes place by continuous growth of the nuclei of the low-temperature polymorph, it is possible to produce a single crystal by allowing the specimen to cool to the transformation temperature progressively along the specimen. The usual requirement is to produce single crystals of the phase that is stable at room temperature. Retention of high-temperature phase is sometimes possible, usually by alloying, which means the phase is no longer pure and hence is not a preferred option.

In crystal growth by phase transformation, there is a transformation front that separates the high- and low-temperature phases, and it is the movement of this front that leads to crystal growth. Phase transitions are usually accompanied by structural change and therefore may involve change in volume (hence density) and many other properties that are relevant to crystal growth. If the nature of the phase transition is diffusive type, with minor symmetry change, the movement of the polycrystalline specimen in the form of a rod through a suitable temperature gradient can result in the single crystal of the low-temperature phase (Figure 7.7a). However, one has to ensure that the rate of traverse does not exceed the velocity of the transformation front.

If the transformation involves major structural changes, every grain of the high-temperature phase will be transformed at least to one grain of the lower phase. Consequently, it will not be possible to obtain a single crystal of the low-temperature phase from polycrystalline high-temperature specimen. The trick here is to obtain a single crystal of high-temperature phase and then again move this material through a suitable temperature gradient so that a coherent transformation takes place, giving a single crystal of the desired low-temperature phase (Figure 7.7b). In this method, single crystalline high-temperature phase is obtained usually by a suitable float zone technique. Melting is done either by electron beam or by induction heating. One can also use optical melting techniques. If all these are done sequentially in a

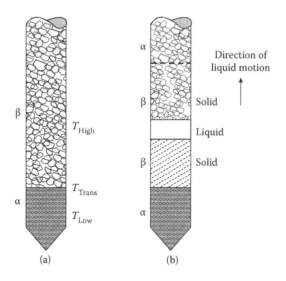

FIGURE 7.7
Growth by solid-state phase transformation with (a) minor and (b) major structural changes.

single run, then there are several phase fronts running through the specimen, one following the other, and it is important to synchronize their velocities. Low-temperature phases of refractory metals such as zirconium [12] and titanium [13] have been produced in single crystalline form by this method.

An alternative approach that has also been tried to prepare low-temperature phase single crystals is subjecting the experimental specimens to thermal cycling. Apparently, the strain produced by the phase change stimulates the growth; hence, the method is akin to the strain annealing technique. The method involves several thermal cycles across the transition temperature T_c by holding the samples for several hours below and above T_c, which is followed by a final annealing at a temperature just below the transition temperature. This method has been applied to obtain single crystals of titanium [14], zirconium [15,16], and also certain rare earth crystals [17].

General Conclusions

From the previous discussion, it is evident that the solid–solid growth methods have largely been applied to refractory metals and alloys and certain ceramics. Although many advantages are attributed to solid–solid growth, because of several disadvantages associated with it, the use of these solid–solid growth methods is limited. Principal among them is the control over orientation of the grown crystal. Cross section of the specimen is often limited. Growth resulting in twinned crystals is also a problem, particularly in face-centered cubic structures. Consequently, these techniques are not exploited extensively by the industry.

References

1. Humphreys, F. J., and M. Hatherly. 1995. *Recrystallization and related annealing phenomena*. Oxford, UK: Elsevier.
2. Aust, K. T. 1963. Large crystals grown by recrystallization. In *Art and science of growing crystals*, edited by J. J. Gilman. New York, NY: John Wiley, pp. 452–478.
3. In-Kwon, J., K. Doh-Yeon, Z. G. Khim, and S. J. Kwon. 1989. Exaggerated grain growth during the sintering of Y·Ba·Cu·O superconducting ceramics. *Materials Letters* 8: 91–94.
4. Yoo, Y.-S., M.-K. Kang, J.-H. Han, H. Kim, and D.-Y. Kim. 1997. Fabrication of $BaTiO_3$ single crystals by using the exaggerated grain growth method. *Journal of European Ceramic Society* 17: 1725–1727.
5. Smallman, R. E. 1970. *Modern physical metallurgy*, 3rd ed. London, UK: Butterworth, p. 393.
6. Andrade, E. N. da C. 1937. Preparation of single crystal wires of metals of high melting point. *Proceedings of the Royal Society A* 163: 16–18.
7. Laudise, R. A. 1970. *The growth of single crystals*. New York, NY: Prentice Hall.
8. Bailey, D. J., and E. G. Brewer. 1975. Improved strain-anneal crystal growth technique. *Metallurgical Transactions A* 6: 403–408.
9. Seraphim, D. P., J. I. Budnick, and N. B. Ittner. 1960. Extractive metallurgy division—Single-crystal growth and purification of tantalum. *Transactions of the Metallurgical Society of AIME* 218: 527–534.
10. Digges, T. G., and M. R. Achter. 1964. Growing large single crystals of niobium by the strain-anneal method. *Transactions of the Metallurgical Society of AIME* 230: 1737–1737.
11. Churchman, A. T. 1954. The slip modes of titanium and the effect of purity on their occurrence during tensile deformation of single crystals. *Proceedings of the Royal Society of London A* 226: 216–226.
12. Mills, D., and G. Craig. 1966. Etching dislocations in zirconium. *Journal of Electrochemical Technology* 4: 300.
13. Cass, T. R., R. W. Quinn, and W. R. Spencer. 1968. Growth of hexagonal titanium and titanium-aluminum single crystals. *Journal of Crystal Growth* 2: 413–416.
14. Anderson, E. A., D. C. Jillson, and S. R. Dunbar. 1953. Deformation mechanisms in alpha titanium. *Transactions of the Metallurgical Society of AIME* 197: 1191–1197.
15. Langeron, J. P., and P. Lehr. 1956. Sur la preparation de gros Cristaux de zirconium et la determination de l'orientation des precipitets d'hydrude de zirconium. *Comptus Rendus* 243: 151–154.
16. Langeron, J. P., and P. Lehr. 1958. Preparation de groscristaux de zirconium et determination de l'orientation des precipitesd'hydrure de zirconium. *Revue de Metallurgie* 55: 901.
17. Tonnies, J. J., and K. A. Gschneidner, Jr. 1971. Preparation of lanthanide single crystals; praseodymium, neodymium and lutetium. *Journal of Crystal Growth* 10: 1–5.

8

Growth from the Melts

Introduction

Crystal growth from melt in essence is a monocomponent liquid–solid growth. Any material that is melting congruently or near congruently can be grown by melt techniques. Ideally, in monocomponent liquid–solid growth, since the rate of growth is not diffusion controlled, growth as such can be quite rapid and the purity of the resultant crystals can be quite high. It is essentially growth by controlled freezing; consequently, in comparison to other growth techniques, it is readily controllable. Except for the growth from aqueous solution, it is perhaps the earliest studied technique. Also, in all probability, it is the most exploited commercial process for single crystal growth, and because of this, sophistication and automation have been introduced into the process.

The simplest method of forming crystals by melt–solid equilibrium is by uncontrolled freezing of a melt. In this process, the initial nucleation is random, and the resultant product is generally a fine-grained polycrystalline ingot. Sometimes, however, single crystal grains of appreciable sizes can be identified in the solid matrix. When such large-sized grains are seen, one can infer that single crystal growth of that material can be achieved with relative ease. In all the melt–solid growth techniques, the main effort is directed toward controlling nucleation so that a single nucleus (or at worst a few) will be formed first and act as a seed on which the growth will proceed. This is achieved by a process that is commonly known as directional solidification. Of course, use of a seed crystal simplifies the nucleation control. The prominent techniques that fall under monocomponent liquid–solid growth are as follows:

1. Bridgman–Stockbarger technique
2. Czochralski technique
3. Zone-melting technique
4. Verneuil flame fusion technique
5. Arc fusion growth
6. Growth by skull-melting

In the following sections, we describe and discuss these important techniques and see how nucleation control is achieved in them. We also consider some of the related developments that have taken place in these techniques.

Bridgman–Stockbarger and Related Techniques

General Considerations

This technique was originally developed by Bridgman in 1925 to grow single crystals of certain metals such as tungsten, antimony, and bismuth [1], and later the method was exploited by Stockbarger to grow large crystals of lithium fluoride [2,3] and hence the name. However, in recent years, it is mainly called by the name of its original inventor. As with all the growth techniques, since the design of the growth apparatus is material specific, the basic technique has been modified by various workers depending on the specific conditions required to grow the chosen crystals. The range of materials grown by the Bridgman technique is so broad that it would not be possible to document all these developments here, and the reader should refer to authentic review papers and books [4,5].

Vertical Bridgman Method

In the Bridgman–Stockbarger technique, the nucleation is produced on a single solid–liquid interface by carrying out the crystallization in a temperature gradient. In the vertical configuration, the material to be crystallized is usually contained in a cylindrical crucible, which is lowered through a temperature gradient in a vertical furnace. In some cases, the crucible is held stationary in a furnace with a temperature profile that produces approximately a linear gradient, and the furnace is then moved upward. In either case, an isotherm normal to the axis of the crucible is made to move through the crucible slowly so that the melt–solid interface follows it. Initially, the entire material in the crucible is melted and homogenized. When the solidification starts, if the crucible has a flat bottom, the first nucleation will be several crystallites at the bottom of the crucible. However, by suitably modifying the shape of the crucible bottom, it is possible to cause one of these crystallites to dominate the solid–liquid interface. Figure 8.1 shows some typical (but not exhaustive) crucible shapes that have been used to restrict the number of nucleation centers so that only one of the crystallites dominates and grows further. Among these, the shape shown in Figure 8.1a is the simplest of the designs wherein the bottom of the crucible is conically shaped with a tip so that initially the melt solidifies only in a small volume. Hence, only one or at worst a few nuclei are formed. This is the most commonly used crucible shape. The shapes shown in Figure 8.1b through e are all intended

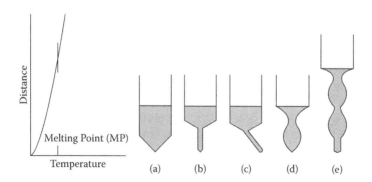

FIGURE 8.1
Temperature profile of a gradient furnace along with (a–e) various crucible tip shapes.

to make sure that only one nucleus survives during the competition and advances to the wider portion of the crucible. For example, the capillary at the end of the crucible increases the chance of one nucleus propagating to the crucible even if initially several crystallites are formed. If the capillary is bent, it promotes the advancement of a crystallite that is favorably oriented. Flared regions in the capillary further enhance the competition between the advancing grains so that in the end only one survives.

In some cases, it is advantageous to cool the conical tip or the end of the capillary of the crucible. Heat leaks may be provided by placing one end of a conductive material in contact with the bottom of the crucible and the other end at the cooler region of the furnace or even projecting it outside the furnace. Spot cooling may also be performed by directing cool gases against the desired region.

Although, in principle, it is possible to use a seed crystal in the Bridgman technique and thereby avoid random nucleation, this is often experimentally tedious because in conventional Bridgman apparatus temperatures are not controlled to such a high degree to prevent the seed from melting. Further, since the crucible materials as well as the furnace tubes used at high temperatures are generally opaque, the visual observation of the melt–solid interface is not possible. However, if the growth occurs at relatively low temperature so that one can use glass or silica ampoules and the furnace is transparent, the seeding process will be less tedious.

Vertical Gradient Freeze Technique

It is also possible to grow crystals without moving either the crucible or the furnace, and this technique is called the vertical gradient freeze (VGF) technique. In this technique, the crucible is positioned in the furnace, and a monotonically increasing temperature profile is applied to the crucible. To begin with, the power to the furnace is brought to a point where the entire charge is melted. If a seed has been introduced in the crucible, one should ensure that only a small portion of the seed is melted to have a homogeneous liquid–solid interface. Once the steady state is reached, the power to the furnace is ramped down in a regulated manner so that the melt starts solidifying from the seed end. There are no moving parts in this arrangement. However, the system permits the rotation of the crucible to average out radial temperature asymmetries if necessary. This technique is particularly suitable in cases such as growth in high-pressure autoclaves where it is difficult to introduce translational motion. It is also important to design the gradient freeze growth system in such a way that the axial temperature profiles remain as parallel as possible over the range of furnace input power required to grow the entire crystal.

Horizontal Bridgman Method

The horizontal Bridgman method has also been developed to avoid the mechanical strains in the crystals caused by the differential thermal expansion between the crystal and crucible material. In this case, the furnace axis is horizontal rather than vertical. Here, the charge contained in an open boat is placed within a horizontal furnace tube, and the directional solidification is caused by slowly withdrawing the boat from within the furnace. Alternatively, the crucible can be held stationary, and the furnace can be traversed. Open boats used in horizontal configuration also facilitate the seeding process. This is done by giving proper shape to the boat at the nucleating end. However, the horizontal method does not give crystals with circular cross-section because in this case the charge has a free surface and does not fill the crucible. Figure 8.2 is the schematic of a horizontal Bridgman setup in which the boat containing the charge is kept inside an evacuated and sealed quartz envelope.

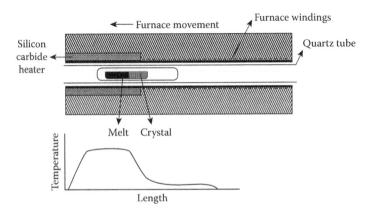

FIGURE 8.2
Horizontal Bridgman setup and the associated thermal profile.

Equipment

From the basic knowledge of Bridgman growth technique, it is clear that there are three principal components in a conventional Bridgman setup: (1) a furnace, (2) a crucible, and (3) an arrangement for mechanical movements. Some general features of these are given in the following sections.

Furnaces

Bridgman furnaces are generally gradient furnaces. In a resistance wire-wound furnace, if the windings are evenly spaced, the hottest region is in the middle of the furnace and the temperature gradually decreases as we approach the ends (Figure 8.3). In such a furnace, the lower region has approximately a linear gradient and can normally be used for the growth purpose. When the crucible is being lowered through the furnace, directional solidification takes place by the upward movement of melt–solid interface to give the crystal. However, in a furnace like this, which is called a single-zone furnace, the grown crystal still resides in a temperature gradient. It is usually advantageous to have two isothermal regions with a gradient between them. This permits annealing of the crystals after growth in an isothermal environment. Such a furnace setup should ideally consist of two furnace regions that are independently controlled with a minimum of thermal coupling between them. For effective thermal isolation, a thermal barrier should separate them. A reflecting baffle (often of Pt sheet) whose size just permits the passage of the crucible is often used to separate the zones.

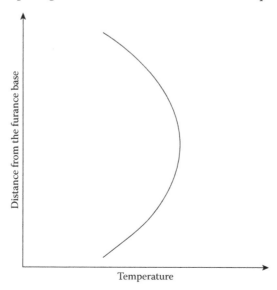

FIGURE 8.3
Temperature profile of a tubular furnace with evenly wound heating wire.

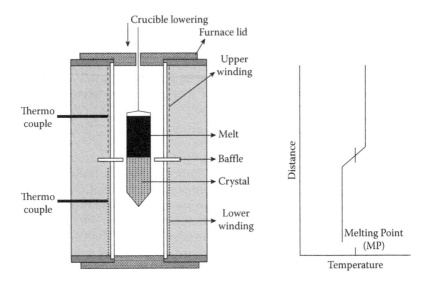

FIGURE 8.4
Schematic of a two-zone furnace consisting of two independent furnace windings separated by a baffle along with the temperature profile.

To ensure isothermal conditions in the separate regions, it is helpful if the inner furnace walls are made of a material of high thermal conductivity. Figure 8.4 schematically represents a typical two-zone vertical Bridgman furnace along with its temperature profile.

A common difficulty one encounters in Bridgman–Stockbarger growth is when there is too small a thermal gradient along the crucible. Many melts supercool appreciably before nucleating. If the melt can supercool sufficiently and the thermal gradient is relatively small, often the whole charge may be below the melting point before the first solid nucleates. When nucleation starts under these conditions, it will occur everywhere in the melt and the growth through the rest of the melt is very rapid. The result will be a polycrystalline ingot with many grains. In such cases, large thermal gradients should be imposed so that the initial nucleation will begin before the entire sample is below the melting point. This can be easily achieved in two-zone furnaces just described. Since the two zones in them are independently controlled, any desired thermal gradient could be set up between them. Also, if the baffle between the two zones is water-cooled, a very large thermal gradient can be set up.

Electrodynamic Gradient Freeze Furnace

With the advancement in technology, sophisticated furnaces have been developed that are used to grow "difficult-to-grow" crystals by directional solidification. One such furnace is the electrodynamic gradient freeze (EDGF) furnace.

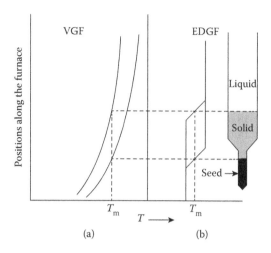

FIGURE 8.5
Axial temperature profiles for (a) vertical gradient freeze (VGF) process and (b) electrodynamic gradient freeze (EDGF) process.

Figure 8.5 schematically compares the thermal profiles in VGF process with the EDGF multizone process. In the VGF approach, the hot zone has a characteristic temperature profile, and as the growth proceeds from the seed into the melt, the entire profile is shifted to lower temperatures.

In the case of EDGF growth, a small segment of the gradient is programmed to pass through an otherwise isothermal melt. There is no need to move either crucible or furnace, and the smoothness of the moving gradient is limited only by the zone resolution. To maintain as much flexibility as possible, the entire furnace is constructed from a large number of identical EDGF units stacked one over the other. In principle, each of these operating units can be used as an independent zone. However, in practice, the total number of zones is considerably reduced by clubbing them appropriately. Since the furnace is of modular design, it provides enormous flexibility in terms of reconfiguring the temperature profile.

Crucibles

Containers, commonly called crucibles, are essential in the Bridgman–Stockbarger technique. The general requirements of a good crucible material are that (1) it should be chemically inert to the charge when it is in molten state so that the crucible is safe and it does not contaminate the charge; (2) it must have a smaller thermal expansion coefficient than that of the crystal since the crystal is grown into it; and (3) its thermal conductivity should be compatible with the heat flow requirements. The inner walls of the crucible should be smooth so that no secondary nucleation occurs on the surface. It is also helpful if the grown crystal does not stick to the crucible because this helps to minimize strains. Also, removal of the grown crystal would be easy. The sticking problem can, however, be mitigated by coating the inner walls of the crucible with suitable materials. Coating may also help in minimizing the effect of differential expansion. Usual coating materials are Aquadag, pyrolized carbon, evaporated silica, Apiezon oil, and so on.

Crucibles used in the Bridgman–Stockbarger technique have been made of glass, vitreous silica, vitreous carbon, alumina, noble metals, graphite, and many other materials. Softening points of glass and vitreous silica being ~600°C and ~1200°C, respectively, they can be used to grow low to moderate melting point materials. However, since they are transparent, in a suitably designed transparent furnace they permit visual observation of the growth process. Split crucibles can be made from these materials for easy removal of the grown crystals, but breaking the crucible often is an easy option because they are relatively inexpensive. As a material, graphite can be easily machined to form split crucibles. However, it should be used only in nonoxidizing atmospheres.

If the material wets the crucible, strain may be introduced in the crystal during extraction. To prevent this, sometimes crucibles called soft molds are used. One version of soft mold is a very thin and easily deformable container from noble metals such as platinum, gold, and silver. Even though such molds are wet by the melt, they introduce no strain in the crystal because they easily deform during cooling to take care of the differential thermal contraction between the crystal and the crucible. If such a mold is so thin that it is incapable of supporting the melt, it may be supported by a stronger and less expensive external crucible. The space between the two containers may be loosely packed with a suitable material like alumina. Soft molds are also made by forming the desired shape in a powder that is packed in a cylindrical container. Magnesia, alumina, and graphite are the common choices for the powders to form such molds.

A crucible for horizontal growth can easily be fabricated. The common shape of the crucible for horizontal growth is boat like. A simple flat plate of a suitable material can also act as a container on which the charge is held together by surface tension. However, the size of the crystal that can be grown with this arrangement would be limited.

Mechanical Movement

A number of methods are available to cause slow motion of the crucible through the furnace. The most common method of lowering the crucible is to attach it to a properly guided wire or a chain, which in turn is wound over a drum fixed to the shaft of a clock motor. Rotation of the motor shaft releases the wire uniformly, thereby lowering the crucible. Even with this type of a simple arrangement, the crucible lowering rate would be within ±0.1% of the set value over a period of several hours, which is quite sufficient in many cases. Lowering rates can be varied by simply changing the diameter of the drum.

Better control can be obtained by mounting the apparatus using antivibration mounts so that vibrations are not transmitted to the crucible. Translating the crucible through a rigid support such as a rod or a pedestal attached to the crucible also tends to eliminate accidental crucible motion. If a metal rod is attached to the crucible bottom, it also facilitates heat leak, which is sometimes desired. Otherwise, one can use a heat-resistant ceramic rod. Sophisticated translation equipment is also available commercially, which can be used for lowering the crucibles.

Any crystal growth process can be refined in many ways. For example, for volatile material the obvious method is to seal the ampoule to contain the melt. An alternate method is to encapsulate the melt with a nonmiscible liquid (e.g., B_2O_3; metal halides; silicone fluids applicable only for low melting point materials, i.e., <400°C).

Melt–Solid Interface Shape

Although interface shapes in Bridgman growth may be complex, the most common shape is parabolic. The interface shape is described by the terms convex, planar, or concave with respect to either melt or crystal, and in Figure 8.6, it is defined with respect to the crystal. If the inner radius of the ampoule is r and the height of the parabola representing the interface is h, then one defines a parameter convexity h_c as

$$h_c = h/r \qquad (8.1)$$

If h_c is positive, the interface shape is convex; if it is negative, it will be concave. Zero convexity represents a planar interface. In Figure 8.6, the direction of heat flow at the melt–solid interface is shown for convex, planar, and concave shapes. Here, K is the thermal conductivity, G is the thermal gradient,

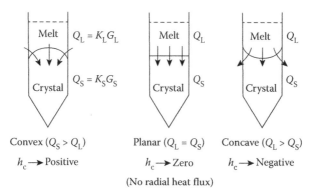

FIGURE 8.6
Melt–solid interface shapes m, in an ampoule of the vertical Bridgman method.

and Q is the quantity of heat. The suffixes L and S refer to liquid and solid, respectively. The conditions for obtaining the respective shapes are also indicated below in Figure 8.6. It should be noted that this diagram is applicable to stationary ampoule. However, in actual growth, the ampoule is traversing with a certain velocity (V).

Ideally, one would like to have a planar interface between the melt and the crystal during growth for the following reasons. Concave interface leads to poor grain selection because any nucleus generated at the wall of the container will grow into the boule. Although grain selection is favored by the convex interface, such shapes often generate twins in the grown crystal; hence, the planar interface is the preferred one. Also, when segregation is considered important, flat interface gives the best radial uniformity. Further, thermal stresses that can generate dislocations will be the lowest for planar interface.

It has been found empirically in many cases that for certain combinations of G and V one can get planar interfaces during growth [6,7]. A plot of G versus V gives a straight line over a range of G and V values, implying that for a particular system the ratio G/V giving planar interface is a constant (Figure 8.7). It should, however, be noted that many other factors such as wall thickness, cone angle, and thermal conductivity of the ampoule do influence the shape of the interface, although not to the same extent as G and V.

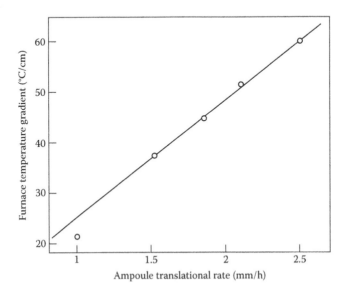

FIGURE 8.7
Furnace temperature gradient versus ampoule lowering rate for planar melt–solid interface for GaSb. (Reprinted from *Journal of Crystal Growth*, 141, Dutta, P. S. et al., Growth of gallium antimonide by vertical Bridgman technique with planar crystal-melt interface, 44–50, 1994, with permission from Elsevier.)

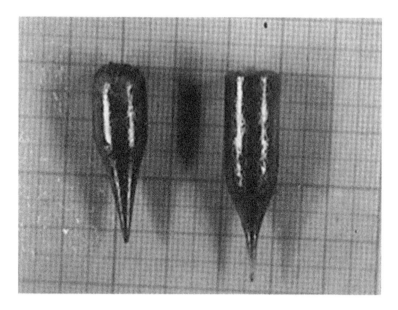

FIGURE 8.8
GaInSb crystals grown with convex and nearly flat interfaces by the Bridgman method. (Reprinted from *Journal of Crystal Growth*, 203, Udayashankar, N. K. et al., The influence of temperature gradient and lowering speed on the melt–solid interface shape of $Ga_xIn_{1-x}Sb$ alloy crystals grown by vertical Bridgman technique, 333–339, 1999, with permission from Elsevier.)

Figure 8.8 [7] shows two crystals of GaInSb; one is grown with convex shape, and the other is grown with nearly planar interface shape.

In the past, demarcation experiments using radiotracers [8], electrical pulsing [9], and thermal pulsing [10] have been carried out to observe and study the liquid–solid interface. In-situ visualization of the interface has also been carried out that enabled one to estimate the extent of supercooling [11].

In the thermal pulsing method, one rapidly quenches the crystal during growth to freeze the interface for a short period and then grows it again at a normal rate. Subsequently, longitudinally cut crystal is etched by a suitable etchant and observed under microscope to see these demarcations. Quenching can be effected by simply switching off the furnace followed by turning it on after a short delay. This can be repeated several times to study the reproducibility and fidelity of the process in demarcating the interface. Typical demarcation lines produced as just discussed are seen in Figure 8.9 [12].

Examples

The Bridgman–Stockbarger growth technique has been used extensively to grow metals, semiconductors, and alkali and alkaline earth halides. By far the largest commercial application of the method seems to be in the growth of

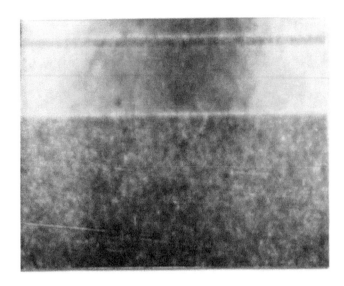

FIGURE 8.9
Demarcations produced in InSb crystal. (From Venkataraghavan, R., et al., *Journal of Physics D: Applied Physics*, 30, L61–L63, 1997. With permission.)

TABLE 8.1

Typical Crystals Grown by the Bridgman Technique

Metals	Semiconductors	Alkaline Earth Halides
Bi	GaSb	CaF_2
Zn	InSb	SrF_2
Cu	GaAs	LiF

alkali and alkaline earth halides. During the early years of the technique, the metals were the main target. Stockbarger's pioneering work on the growth of large crystals of LiF and CaF_2 provided a basis for an extensive commercial growth of halide crystals for optical purposes. In all the aforementioned work, the vertical configuration of the apparatus was used. Semiconductor crystals have been grown in both the vertical and horizontal configuration. The list under each is exhaustive, and Table 8.1 provides only a representative list.

A number of organic crystals have also been grown in recent years by the Bridgman technique.

Growth of Indium Antimonide

As a case study, we describe the growth of indium antimonide crystals by the Bridgman method in some detail. In fact, antimonides of Ga and In have been investigated extensively in recent years in terms of both growth and physical properties [13–15]. Indium antimonide melts at ~527°C; therefore, one can use silica crucible for growth. However, since the vapor pressure of Sb at the growth temperatures is very high, open crucibles cannot be used. In the author's laboratory, they were grown both by vertical and horizontal methods in sealed crucibles. While a BCG365 crystal growth system from Metals Research (UK) was used for vertical configuration, a homemade apparatus was used to grow crystals in horizontal configuration.

Materials Synthesis

Indium antimonide is synthesized from high-purity elements taken in appropriate quantities using a mixing furnace. The mixing furnace is a horizontal cylindrical furnace with a 45 mm ID, 50 mm long muffle wound with nichrome wire. The furnace can be powered to 250 watts to reach a temperature of ~850°C to give an 8 cm long isothermal region. The furnace temperature is controlled through a temperature controller having an accuracy of ±1°C. A stainless steel tube of 25 mm ID, 28 mm OD, and 64 cm long coaxial to the furnace muffle is supported by bearings on both sides. The quartz ampoule filled with the starting elements is placed horizontally inside this tube. The steel tube has the capability of periodic rotation both in clockwise and anticlockwise directions at a rate of nine rotations per minute facilitating proper mixing of the molten elements to form the compound. Synthesis of InSb involves the preparation of homogenized sample of InSb from elemental In and Sb taken in 1:1 atomic ratio. Since the atomic weights of In and Sb are 114.82 and 121.75 amu, respectively, the ratio by weight is 1:1.06. Initially, before weighing, elemental In is etched with HNO_3 and Sb with aqua regia for the removal of their surface oxides. Then, a chemically cleaned semiconductor-grade (GE214) silica tube of 16 mm bore and 8 cm in length with a round bottom filled with the required starting elements is sealed under a vacuum of ~10^{-6} Torr. The filled ampoule is then loaded into the synthesizing unit, and mixing is carried out for 24 h. The synthesized material is removed from the ampoule after cooling the furnace gradually to room temperature. Generally, the starting material for a required number of growth runs should be synthesized together to eliminate batch-to-batch variation in stoichiometry. Extreme care should be taken in the transfer of material to maintain the accuracy to the level of µg/g of material.

Crystal Growth

Vertical Bridgman Growth

A schematic of the furnace and the crucible system of BCG 365 crystal growth unit is shown in Figure 8.10 [6] along with a typical temperature profile. The complete system is shown in Figure 8.11. In this unit, the charge is melted in a single-zone tubular furnace, 21 cm long and 2.5 cm bore. This single-zone furnace is contained within a cylindrical double-walled, water-cooled, and

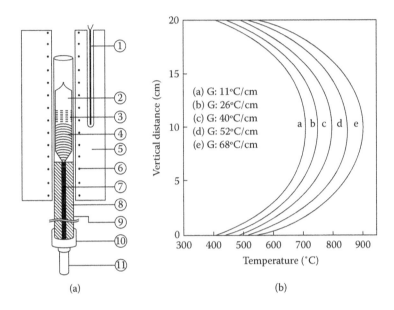

FIGURE 8.10
(a) Schematic of the furnace and crucible assembly: (1) NiCr/NiAl thermocouple, (2) Quartz ampoule, (3) Melt, (4) Crystal, (5) Furnace, (6) Furnace element, (7) Ampoule stem, (8) Quartz wool, (9) Quartz tube, (10) Stainless steel coupler, (11) Crucible lowering rod. (b) temperature profiles of the furnace for different input powers. (Reprinted from *Journal of Crystal Growth*, 141, Dutta, P. S. et al., Growth of gallium antimonide by vertical Bridgman technique with planar crystal-melt interface, 44–50, 1994, with permission from Elsevier.)

transparent growth chamber assembly that is mounted on a mainframe plinth. A hinged radiation shield isolates the growth chamber for protection. The growth chamber is evacuated to 10^{-3} Torr throughout the experiment in order to avoid connective losses and to reduce convection-driven instabilities. The temperature of the environment is controlled with an accuracy of ±0.1°C by a Eurotherm proportional band controller configured to the system. The thermal profiling of the furnace is precisely mapped (Figure 8.10), and gradients in the range of 10°C–70°C/cm at the melting point of InSb materials could be imposed by varying the input power to the furnace. The ampoule could be lowered through the chosen temperature gradient of the furnace with selected translational rates in the range of 1–25 mm/h. To grow InSb single crystal, the synthesized material is removed from the quartz ampoule and then subjected to proper chemical cleaning and then filled in 10 mm bore and 12 cm long silica ampoule drawn from semiconductor-grade GE214 tube. The ampoule is provided with a proper conical tip at the bottom to facilitate nucleation for single crystal growth. The filled ampoule is then sealed under dynamic vacuum of ~10^{-6} Torr and later loaded into the growth unit. The furnace temperature is increased to 750°C (a typical run), and the melt is homogenized for 3 h. The growth is carried out at different thermal gradients.

FIGURE 8.11
Metals Research BCG 365 crystal growth unit.

The lowering rates selected were in the range of 1.5–2.5 mm/h. Even after solidification, the lowering rate was maintained the same until the entire ampoule was out of the furnace so as to obtain an effective annealing of the crystal. The ingot was taken out by carefully cutting open the quartz ampoule. Crystals with a maximum diameter of 20 mm could be grown with this unit. A typical crystal grown is shown in Figure 8.12.

Horizontal Bridgman Growth

Figure 8.13 shows a horizontal Bridgman system in which the crucible is held stationary and the furnace is translated. The growth system consists of a single-zone furnace through which a 1 m long silica tube with a bore of 20 mm passes coaxially. The quartz ampoule filled with the charge is placed horizontally inside this tube. The travel speed of the furnace is selected by coupling the lead screw to a turret-type commercial geared motor capable of angular speeds

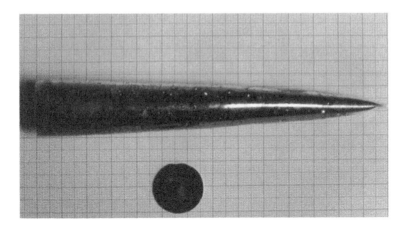

FIGURE 8.12
Bridgman grown InSb crystal and a wafer cut from it.

FIGURE 8.13
Photograph of a horizontal Bridgman setup.

from 0.00082888 to 10.35 rpm with 20 intermediate steps. The fabricated furnace, coaxial to the silica tube, has a bore of 31 mm and is capable of maintaining temperatures up to ~1000°C at its center. The furnace temperature is monitored using proportional-integral-differential (PID) (Eurotherm) temperature controller, which is compatible with a type-K thermocouple. The temperature profile of the furnace is similar to the one shown for a vertical Bridgman furnace with maximum temperature being at the midpoint in the furnace.

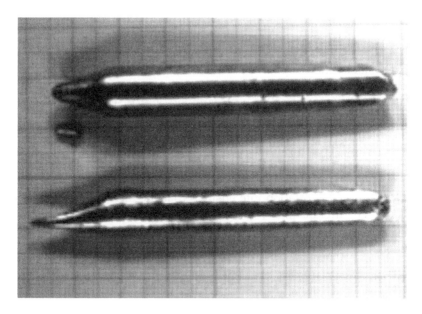

FIGURE 8.14
Typical crystals of InSb grown by the horizontal Bridgman method. (With kind permission from Springer *Bulletin of Materials Sciences*, Growth and characterization of indium antimonide and gallium antimonide crystals, 24, 2001, 445–453, Udayashankar, N. K., and H. L. Bhat.)

For the growth of single crystals of indium antimonide using the horizontal Bridgman technique, the chemically synthesized InSb polycrystalline material is transferred to a quartz ampoule having a conical tapering and then evacuated (~10^{-6} Torr) and sealed. The ampoule is then loaded into the horizontal Bridgman unit and given a small inclination of about 10° to ensure proper filling of the conical tip. The furnace temperature is increased to 750°C and homogenized for 3 h. The temperature gradients in the range of 35°C–85°C/cm and furnace travel rates in the range of 1–5 mm/h are employed for the growth. Typical crystals grown by this method can be seen in Figure 8.14 [15].

Limitations

As mentioned earlier, one of the main problems in Bridgman growth is that the growth occurs by self-nucleation. As a result, the crystallographic orientation of the grown crystal is not known. However, if the crystal has an easy growth axis, this axis might coincide with the ampoule axis. Although some efforts have been made to use seed crystals in Bridgman growth, this is not very common, mainly because the seed melt interface is not visible. Hence, in most cases, the orientation of the crystal has to be determined after each growth. Also adhesion of the crystal with the ampoule often poses problems related to crystal extraction and thermal stresses. Both these problems are overcome in the Czochralski technique, which we discuss in the next section.

One other problem with a Bridgman-type system is the lack of control over stirring. In the case of the horizontal Bridgman technique, horizontal temperature gradient facilitates some convective mixing. However, in conventional vertical systems, the temperature increases upward and does not promote convection. Even though, in principle, mechanical stirring is possible in the case of open ampoules, it is quite cumbersome. Instead, one can use the accelerated crucible rotation technique commonly abbreviated as ACRT. In this technique, the crucible is rotated about its vertical axis in the clockwise direction with increasing rotational velocity to accelerate. On reaching the maximum velocity, the crucible rotation is decelerated to come to rest and the whole cycle repeats in the counterclockwise direction. The continuous variation in crucible rotation rate ensures that the melt and crucible rotate transiently at different rates leading to flow patterns which cause mixing. Flow patterns in Bridgman ampoules are quite complicated and depend on a number of parameters.

Czochralski and Related Techniques

The method that has been used to grow a wide range of materials from low melting point metal such as Ga (29.8°C) to high melting point oxide spinel (2135°C) is the pulling technique due originally to Czochralski [16]. This method has become very popular largely because of the rapidity with which it can produce relatively large crystals that are highly perfect. The principal advantage of crystal pulling is that the growth can be initiated on a seed under controlled conditions which arises because both the seed and the crystal are visible during growth and the crystal grower can use this facility to guide him or her to adjust various process parameters to obtain crystals with a high degree of perfection. In addition, growth in any direction can be performed when oriented seeds are available. The method also permits the addition of suitable dopants during growth. Even though it is necessary to have a crucible to melt and hold the charge that may often contaminate the charge and hence the crystal, the growing crystal is free from physical constraints imposed by the crucible. This is mainly because the growing crystal is not in contact with the crucible.

In principle, any material with a congruent melting point can be grown by the pulling technique. Other desirable properties are high thermal conductivity, low vapor pressure, low viscosity, suitable growth habit, absence of easy cleavage, and no phase changes between the melting point of the material and room temperature. A material without the congruent melting point will have to be grown by some other method. However, lack of any of the other properties mentioned earlier can usually be overcome by some ingenious methods.

The elements of the Czochralski technique are illustrated in Figure 8.15a. The charge material is contained in a crucible, which can be heated to above its melting point. A pull rod with a chuck containing a seed crystal at its lower end is positioned above the crucible. The seed crystal is dipped into the melt whose temperature is adjusted until a meniscus is supported by the seed crystal. The pull rod is then slowly lifted at a desirable rate so that growth takes place without the detachment of the seed. When the seed is not available, nucleation can be initiated on a twisted platinum wire. The polycrystalline mass thus formed on the wire is pulled for a while and then its diameter is reduced (a procedure known as necking down) by slightly raising the furnace temperature. Once the crystal is necked down, a single crystallite will usually dominate the growing interface. The diameter of the crystal can then be increased to get a single crystal of the desired cross-section. Usually the seed is rotated by rotating the pull rod at a certain rate that changes from material to material. In addition to the rotation of the seed, the crucible can also be rotated, which will ensure symmetrical heating of the melt. Because of these, the grown crystal will normally be cylindrical unless the crystal has a very strong tendency toward faceting.

A simpler technique that is somewhat related to Czochralski growth is the Kyropoulos technique [17,18]. In this technique, the seed is inserted into the melt contained in an appropriate crucible. However, the seed is not withdrawn from the melt but grown into it. The growth is achieved by causing an isotherm corresponding to the melting point of the charge to move from the seed downward into the crucible (Figure 8.15b). This can be simply done by cooling the seed via the seed holder, which establishes an effective heat leak from the furnace. Alternatively, the crucible can be moved downward through a thermal gradient. Consequently, Kyropoulos equipment is simpler (because pulling is not required). However, with this method, there are problems associated with controlling the shape of the crystal because the growing crystal tends to assume its natural habit. Also, when the crystal is finally removed from the melt, it would suffer a thermal shock.

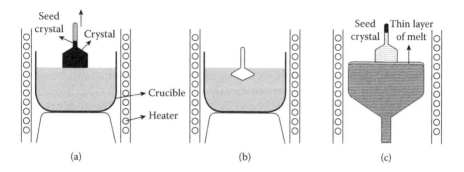

FIGURE 8.15
Schematic of (a) Czochralski, (b) Kyropoulos, and (c) pedestal techniques.

There is also a third variant called the pedestal method (Figure 8.15c) in which the crystal is pulled from a shallow melt formed at the end of a charge rod of larger diameter. Since there is no crucible here, there is no contamination due to the crucible. The technique has been improvised a great deal in recent times by incorporating focused laser beams to melt the material, and this version is being called the laser-heated pedestal growth technique. The pedestal method is very close to the zone-refining technique to be discussed later.

Equipment

The essential features of a modern and versatile crystal pulling apparatus are as follows:

1. A rigid vibration-free mainframe.
2. A pull head containing a precision lead screw and gear system driven by synchronous motors to provide a wide range of pulling and rotation rates to which the pull rod is attached (translation units made from Cyberstar, France, for example, are capable of giving pulling speeds from 0.01 mm/h to 99.99 mm/min). Rotation rates are commonly in the range of 1–200 rpm. The pull head often contains certain sensors that could maintain a constant diameter through a feedback mechanism.
3. A range of crucibles to hold the melt.
4. A heating system with a stable power supply to heat the crucible to the required temperature and an accurate temperature-controlling mechanism.
5. A variety of work chambers ranging from a simple fused silica envelope to a water-cooled high-pressure vessel suitable for high-pressure growth.
6. A facility to create various ambience in the growth chamber. Such a facility would have gas metering and purification systems, vacuum pumps, and gauges.
7. An CCD camera and monitor to view the growing crystal.

A much simpler version of a crystal pulling apparatus that can be operated at ambient pressure is schematically shown in Figure 8.16. It consists of a rigid frame to which a pull head is attached. The pull head contains a pull rod and a load cell, which senses the weight of the crystal grown. At the base of the unit rests a resistively heated tubular furnace. Inside the furnace a crucible is kept on a pedestal. The furnace lid has a viewing port. With an associated temperature controller that facilitates ramp up, hold, and ramp down at the required rate, the system can be used to grow a fairly large number of crystals.

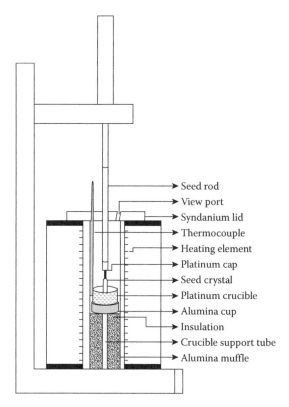

Seed rod
View port
Syndanium lid
Thermocouple
Heating element
Platinum cap
Seed crystal
Platinum crucible
Alumina cup
Insulation
Crucible support tube
Alumina muffle

FIGURE 8.16
Essential components of a Czochralski puller.

Crucibles

Platinum crucibles are used in the case of many oxides whose melting points are below that of platinum. In the case of silicon, the molten silicon is contained in a silica crucible. Since the solubility of carbon in germanium (melting point: 937°C) is practically zero at the melt temperature, germanium can be melted in a graphite crucible. Materials such as yttrium iron garnet, sapphire, ruby, calcium tungstate, and spinel can be pulled from iridium crucibles. Other common crucible materials are gold, Rh, and vitreous carbon. Organic crystals with low melting points can be grown using glass crucibles.

If the growing crystals are pulled to cold ambience, they will experience severe thermal stresses. To minimize this, crystals are pulled into after heaters. After heaters typically consist of an inverted platinum crucible or an iridium-lined ceramic crucible with a hole at its base through which the pull rod passes. Heating is provided by coupling the platinum after heater or iridium lining to the heating source.

Heaters

Heating in Czochralski growth systems is either resistive or inductive. Resistive heaters are wire-wound single-zone furnaces, similar to the ones used in a Bridgman technique except that they are usually shorter in the case of Czochralski growth. Induction heating can be used if either the charge or the container couples to the radio frequency field. Alternately, a susceptor surrounding the container can be used to heat the crucible. Since most melts are conducting, once the charge is melted the radio frequency field also couples to the charge. Radio frequency generators (450 kHz) with capacity ranging from 5 to 50 kW are easily available in the market. For insulating materials, high-frequency generators operating at MHz range would be more desirable because it is easier to match the impedance at higher frequencies. The work coils are usually made of copper tube through which water is circulated. The coils are mostly evenly spaced, although graded windings are also used to generate specific temperature gradient.

Diameter Control

Although crystals grown by pulling technique are usually circular in cross-section, special effort has to be put into growing them as cylinders with uniform diameter. Cylinder-shaped crystals with uniform diameter are preferred by the industry because of the underlying economic advantage. In the initial stages of technology development, the diameter was controlled by human skills and experience. Figure 8.17a shows a crystal of sodium chloride

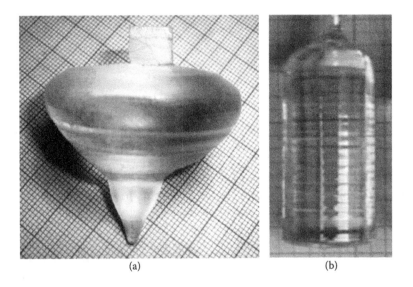

(a)   (b)

FIGURE 8.17
(a) Sodium chloride crystal grown without diameter control and (b) lithium niobate crystal grown with diameter control.

grown by a less experienced crystal grower with manual diameter control. Manual adjustment of the temperature led to detachment of the crystal from the melt, resulting in a top-shaped crystal. Modern commercial pullers are largely equipped with automatic servo controlled systems that maintain a constant diameter.

Automatic diameter control systems are of either mechanical or optical type. In the optical system, one uses the bright meniscus ring that is formed around the crystal where it is in contact with the melt surface. This is caused by the reflection of hot upper regions of the crucible by the meniscus. Any change in the diameter of the growing crystal leads to a lateral shift in the bright ring image, which leads to the change in the output signal of a detector like an optical pyrometer which views it. This signal change can be used as a feedback to carry out appropriate correction in the furnace input power. Alternately, a laser beam focused on the meniscus after reflection can be caught by a photodiode, and the output of the photodiode can be used as a controlling signal.

Optical methods are applicable only if the melt has a good reflectivity like in semiconductor materials. However, oxide melts, in general, have low reflectivity in comparison to compound semiconductors. For materials with low reflectivity, mechanical methods are more appropriate. This method involves the weighing of the crystal during growth. Bardsley et al. [19] were the first to report the diameter control by crystal weighing. Being applicable to almost all kinds of materials, it has now become the preferred choice for a general purpose crystal puller.

The mechanical diameter-controlling systems are commercially called load cells, which contain a set of strain gauges (to which the pull rod is attached) whose combined strain varies linearly with the weight of the crystal being pulled. Thus, load cell works on the basis of differential weight added onto the growing crystal for a fixed diameter; that is, for a constant pull rate, [π (diameter/2)^2 × incremental height × density] should be constant. The actual differential weight is measured and compared with the reference value generated by inputting the appropriate parameters, and the error signal is used to control the heater input power. Figure 8.17b shows a lithium niobate crystal grown with automatic diameter control.

As in the Bridgman growth method, in the Czochralski technique the crystals with high perfection are obtained by maintaining the liquid–solid interface shape flat during the entire growth process. This is achieved by suitably adjusting the radial and axial heat flow conditions so that the isotherms are normal to the growth direction.

Example

Although pulling technique was first applied to metals and then exploited with great success to grow semiconductors, the subsequent development occurred when it was applied to grow refractory oxides. With the development of lasers and nonlinear optics, many laser hosts and crystals with a high degree of optical nonlinearity were grown. The first oxide crystal to be grown was scheelite ($CaWO_4$) [20]. The list of oxide crystals grown by Czochralski technique is continuously expanding.

Growth of Congruent Lithium Niobate Crystals

In this section, we discuss the growth of lithium niobate in some detail as a prototype material. Lithium niobate is an excellent ferroelectric and nonlinear optical material with diverse physical properties and a wide range of applications. As we discussed in Chapter 6, this material melts congruently at composition in which the Li/Nb ratio is 0.94. Commercial crystals are mostly with congruent composition. Initially, we describe the growth of congruent lithium niobate (CLN). Efforts made by researchers to grow stoichiometric variety are discussed in the next section. CLN melts at 1253°C. The starting material can either be commercially available $LiNbO_3$ or that prepared by solid-state synthesis taking $LiCO_3$ and Nb_2O_5 in congruent composition.

A variety of crystal pullers are available commercially. The companies also supply custom-made pullers. Since lithium niobate is grown in a flowing ambient of oxygen + argon mixture at normal pressure, a simple quartz envelope enclosing the growth station is good enough. One such puller purchased from Cambridge Instruments Ltd, UK is shown in Figure 8.18 along with the schematic of the growth station.

A typical procedure to grow the crystal using such a puller is as follows:

1. Load the crucible tightly with the charge and position it in the growth chamber/station.

2. Fix the seed crystal into the seed chuck. Before inserting the seed one should make sure that it is free from processing damage.

3. If the crystal has to be grown in an ambient gas, the growth chamber has to be preferably evacuated first and then flushed with ambient gas. A 1:1 mixture of oxygen and argon gases is the preferred ambient gas for lithium niobate.

4. Melt the charge and hold its temperature a little above the melting point.

5. Bring the seed to a position a little above the melt surface and allow it to attain thermal equilibrium. Apply seed rotation and increase it slowly to its optimum value. Then, lower it very slowly into the melt to minimize thermal shock.

6. Make fine adjustments to the melt power until the seed supports a meniscus.

7. Commence pulling and adjust the temperature such that a thin neck is grown onto the seed crystal.

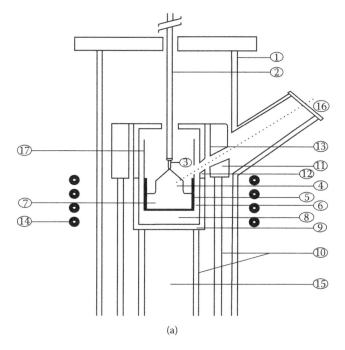

(a)

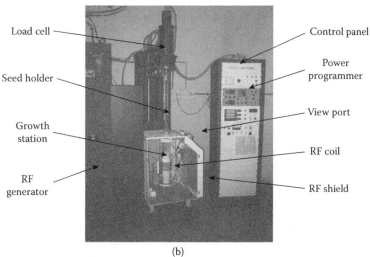

(b)

FIGURE 8.18

(a) Schematic of the growth station: (1) y-type transparent quartz tube, (2) alumina seed rod, (3) platinum seed holder, (4) growing crystal, (5) platinum crucible, (6) zirconia felt, (7) molten charge, (8) zirconia grog, (9) magnesia support crucible, (10) alumina tube, (11) Trion blanket, (12) magnesia inverted cup, (13) zirconia felt, (14) radio frequency coil, (15) zirconia grog, (16) viewing window, and (17) platinum after heater. (b) Autox crystal puller from Cambridge Instruments Ltd, UK.

8. Increase the crystal diameter slowly and smoothly by decreasing the power to the heater until the growing crystal attains the required diameter.

9. During the course of growth, the power to the heater will have to be continuously monitored to maintain a constant diameter to the growing crystal. If the growth unit has an automatic diameter-controlling mechanism, it can be put into action once the required diameter is achieved.

10. Once the required length of the crystal is attained, switch off the automatic diameter-controlling mechanism and slowly increase the power so that the diameter of the crystal reduces. Continue this until the crystal gets detached from the melt. Alternately, the pull rate can be increased to cause the detachment.

11. Lift the crystal a little above the melt into the after heater for annealing.

12. After annealing the crystal for a required duration, cool down the furnace to room temperature.

13. Stop the crystal rotation and extract the crystal by carefully detaching it from the seed chuck.

Figure 8.19 shows typical crystals of congruent $LiNbO_3$ thus grown.

Refined Processes

Since the development of the Czochralski technique, many refinements have taken place in the basic process to make it more versatile. Some of the notable refinements are discussed next.

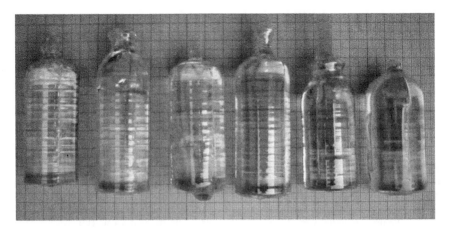

FIGURE 8.19
Typical lithium niobate crystals grown by the Czochralski technique.

Growth of Stoichiometric Lithium Niobate

Since it is a well-studied material, enormous data are available on the properties of CLN [21,22]. In this material, as a result of Li deficiency, there exist intrinsic defects in the form of Nb-antisites and Li vacancies in its lattice, which adversely affect its performance in certain specialized applications. In recent years, it has been realized that $LiNbO_3$ crystals with near-stoichiometric composition have superior properties compared to CLN. For example, stoichiometric lithium niobate (SLN) crystals have reduced threshold electric field for poling to produce periodically poled devices [23]. Laser damage and photorefractive damage thresholds of SLN are enhanced at a much lower level of Mg/Zn doping compared to CLN, which is a positive factor in producing high-quality crystals [24,25]. Because of these and many other advantages, SLN is gaining importance.

SLN crystals are mostly grown by the double-crucible Czochralski method pioneered by Kitamura et al. [26]. In this method, the crystals are pulled by the Czochralski crystal puller from a Li-rich melt (Li_2O 58–60 mol%) in the inner crucible, and the stoichiometric $LiNbO_3$ powder is supplied smoothly and continuously to the melt in the outer crucible. A hole in the side wall of the inner crucible allows the melt from the outer crucible to flow into it to maintain the composition of the melt in the inner crucible more or less constant. Pulling rate and material feeding rate are synchronized such that there is no net loss from the crucible. Crystals as large as 40 mm in diameter and 50–100 mm in length were grown by this group. From the lattice parameter and Curie temperature measurements, the chemical composition throughout the grown crystal was found to be close to the stoichiometric composition.

The other methods used to produce SLN crystals are vapor transport equilibration technique [27], top-seeded solution growth technique with K_2O as flux [28], zone-leveling technique [29], and top-seeded solution growth technique with 58 mol% Li_2O composition [30]. The author's laboratory has used the method similar to the one reported in reference 30 using conventional Czochralski puller with excess Li_2O (58.6 mol%) to produce pure and doped crystals, which are shown in Figure 8.20 [31].

Liquid Encapsulation Technique

Growth of volatile materials such as GaAs and InAs poses their own problems. For example, the dissociation pressure of As over molten GaAs at the melting point is $\approx$1 atm; hence, the melting of GaAs in a conventional crystal puller results in most of the arsenic evaporating off and condensing on the colder wall of the enclosure. Initial attempts to overcome this problem were based on having the As vapor in equilibrium with the melt by maintaining the whole enclosure at a temperature above the sublimation point of the As. This was achieved by having a reservoir of As whose temperature was adjusted to control the vapor pressure. Pulling and rotation were achieved

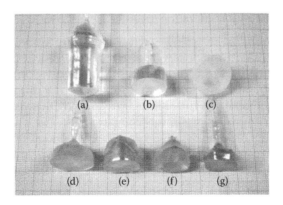

FIGURE 8.20
(a) Nearly stoichiometric pure and (b–g) doped lithium niobate crystals. (Reprinted with permission from [Babu Reddy, J. N., et al., *Journal of Chemical Physics*, 128, 244709, 465–470, 2008], American Institute of Physics. With permission.)

by two methods. One called the syringe method involved a bearing surface of boron nitride, which provided a mechanical seal between the pull head and the chamber capable of working at a temperature >600°C and in an ambient arsenic vapor [32,33]. The other method used the coupling of an external magnet to a high Curie point alloy attached to the pull rod, which was sealed within the heated enclosure [34,35]. Pulling of the magnet led to the pulling of the seed rod.

These cumbersome methods have been largely replaced by what is commonly known as "the liquid encapsulation technique" [36,37]. In this technique, a layer of boric oxide is used to cover the melt (Figure 8.21a). Boric oxide on melting (melting point: 450°C–465°C) becomes a glassy viscous fluid in which As is virtually insoluble. This eliminates almost completely the loss of As from the melt. Initially lumps of source materials are loaded into the crucible over which the encapsulant (boric oxide) is placed. Then, the enclosure is raised to a pressure greater than the dissociation pressure of the charge. When the crucible is subsequently heated, the encapsulant melts first to form a layer around the charge, thereby isolating it. Pulling can then be performed by lowering the seed through the lighter boric oxide layer and the growth occurs at the interface of the two liquids. Furthermore, as the crystal rises, a thin coating of boric oxide remains at its surface, which effectively prevents evaporation of As from solid during cooling (Figure 8.21b). However, for the growth of semiconductors like InP and GaP (dissociation pressure ~15 and ~30 atm, respectively), one has to enclose the charge in high-pressure vessels to prevent phosphorus from escaping in addition to encapsulating. Such high-pressure growth chambers are commercially available. The liquid encapsulation technique has been used in many other cases also.

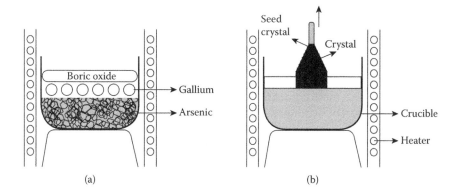

FIGURE 8.21
(a) Schematic of loading of the crucible for growing GaAs using liquid encapsulation technique and (b) pulling through the B_2O_3 liquid layer.

Shaped Crystal Growth

Modern crystal technology uses crystals of various shapes and sizes. Shapes may include plates, rods, tubes, and so on. Usually, the required shapes are obtained by postgrowth processes like cutting, drilling, grinding, milling, and so on. If crystals of such shapes and dimensions can be grown directly, they can be used as final products without applying the aforementioned postgrowth processes. In principle, one can profile the container to the required shape and try to grow crystal into it by melt growth. However, it has been found that this is not practical, in many cases for a number of reasons, some of which were discussed in sections dealing with the Bridgman technique. Interestingly, the crystal pulling technique facilitates the growth of shaped crystals. This was first demonstrated by Stepanov [38,39] and later developed by many. In the literature, these are referred to as the Stepanov method, edge-defined film-fed growth, shaping by capillary action, and so on. In these techniques, as the crystal is not restricted by the crucible walls, its cross-section heavily depends on the growth parameters, which need to be strictly controlled. Any deviations in these result in the changes in crystal cross-section called pinch formation.

Shaped crystals have been grown from semiconductors, metals, and dielectrics. In particular, sapphire ribbons that serve as substrates for integrated circuits, silicon ribbons that are used in solar energy converters, germanium rods that are used in semiconductor devices, and sapphire tubes that are used for illuminating high-pressure sodium lamps and laser devices have been grown by shaped crystal growth methods. By automating the growth system, the weight, shape, and quality of the crystals can be constantly monitored. Readers are encouraged to read an exhaustive chapter written by Tatarchenko in *Handbook of Crystal Growth*, Vol. 2b to know more about the shaped crystal growth [40].

Stepanov Method

The Stepanov method should be considered a natural extension of the Czochralski method, in which deliberate implementation of shaping of the liquid column is carried out. The basis of the Stepanov method is that the shape of the grown crystal is determined primarily by a fixture or a die shaping the melt column. The shaped liquid–melt column is then transformed into solid material via the control of temperature, pull rate, and other growth parameters (Figure 8.22). Different versions of changing the cross-sectional profiles of the growing crystals have been developed based on the Stepanov method. In these techniques, basically the melt meniscus formed is detached from the edges of the first die specifying the initial shape, and then it catches the edges of another shaper specifying the subsequent profile of the cross-section. This is mainly achieved by mass flow rate of the melt supplied to the crystallization front, which is controlled by several means. The control could be achieved by changing the relative position of the different parts of the shaper, change in the position of the shaper with respect to the melt level, sequential use of different shapers, varying the temperature of the crystallizing front, changing pulling rate, and so on.

Several other modifications have been introduced to the Stepanov method. Among them growth by local shaping technique (LST) is worth mentioning. In LST technique, the shaper is kept asymmetric with respect to the axis of the crucible, and by rotating the seed crystal, the growth is achieved layer by layer by slowly pulling the seed. This technique was used to grow lithium fluoride crystals in the form of tubes [41]. By moving the shaper away from the axis of the crucible in a programmed way, sapphire crystals with hemispherical shape have been grown [42]. Successful growth was cleverly achieved by using the shaper with inclined end.

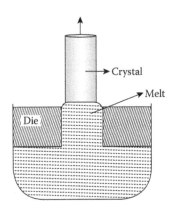

FIGURE 8.22
Principle of the Stepanov method.

Edge-Defined, Film-Fed Growth

Extensive work carried out by La Belle and Mlavsky [43–45] on the growth of sapphire tubes led to the development of what is called edge-defined, film-fed growth (EFG) technique. During the growth experiments, they noticed that if the top surface of the shaper was flat, the molten material spread out to the edge and stopped implying that the shaper is wetted by the melt. Hence, the diameter of the die, not of the melt column, dictated the diameter of the filament produced (Figure 8.23). Consequently, the technique was named "edge-defined film-fed growth." The EFG technique of crystal growth today more or less follows the same process that Labelle ultimately arrived at. For example, to grow shaped sapphire crystals, Al_2O_3 is melted in a molybdenum crucible. The melt "wets" the surface of molybdenum die and moves up by capillary action. A sapphire "seed crystal" of desired shape and quality is dipped into the melt on top of the die and "pulled" out, crystallizing the Al_2O_3 in a shape as determined by the die. Crystal orientation can be tightly controlled—any axis or plane can be produced using proper seeds.

The difference between EFG and Stepanov methods is in what exactly shapes the crystal. In EFG technique, it is the shape of the die, whereas in the Stepanov method, the die shapes the melt column, which solidifies into crystal. Of course, in both cases, it is the surface tensional force that leads to capillary rising.

Micro-Pulling-Down Method

This method developed mostly in the first decade of this century [46,47] is very useful in growing crystal fibers of many materials. The development of

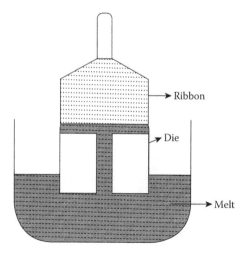

FIGURE 8.23
Edge-defined, film-fed growth method.

this technique is prompted by the advantage of single crystalline fiber lasers over the commercially available double-clad amorphous fiber-based lasers in terms of increasing the core dimension, which in turn allows for attaining large continuous wave power or extremely high peak power pulses.

The schematic of the micro-pulling-down method is shown in Figure 8.24. In this method, the melt (oxide, fluoride, metal) residing in a crucible is transported downward through micro-capillary channel(s) made at the bottom of the crucible. The capillary action and gravity support the delivery of the melt to the liquid–solid interface formed under the crucible. By suitably configuring the crucible bottom, various shapes (fibers, rods, tubes, plates, etc.) and sizes can be obtained. The crystal's cross-section can range from 0.1 to 10 mm.

In a typical growth run, the crucible is first filled with the charge and heated to a temperature so that the starting material is completely melted. At this stage, the melt does not flow down the capillary. Then, the seed crystal is moved upward and slowly brought into contact with the crucible. Subsequently, through appropriate adjustment of crucible temperature and position of the seed crystal, melt is made to touch the seed crystal to form a meniscus. Once a stable interface is formed by establishing an appropriate temperature gradient, growth is commenced by pulling the seed crystal downward. On reaching the desired length of the fiber crystal, the growth run is terminated by separating the as-grown crystal fiber from the meniscus. The system is then cooled down to room temperature.

The main advantage of the pulling-down scheme is that the incorporation of bubbles into the crystals is greatly reduced in comparison to pulling-up methods. The other advantage is that it is relatively easy to feed the melt with

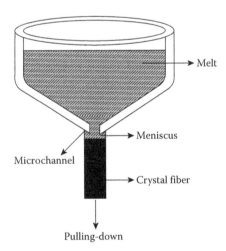

FIGURE 8.24
Schematic of micropulling method.

additional raw materials because the top surface of the melt is generally free and available for feeding.

A large number of scientifically and industrially important crystal fibers have been successfully grown using this method—SiGe, $LiNbO_3$, $KNbO_3$, and $Bi_4Ge_3O_{12}$, to mention just a few.

Zone-Melting Techniques

The zone-melting technique was first invented by Kapitza [48] to grow high-purity bismuth crystals in glass tubes to study their electrical resistivity at very low temperatures under high magnetic fields. Later, in the early 1950s, it was developed as a refining technique by Pfann [49,50] at Bell Telephone Laboratories in response to the demand for extremely pure germanium (melting point: 958.5°C). Purification of germanium was essential for the development of transistor, which in turn led to the revolution in electronics. Subsequently, it was applied to silicon (melting point: 1420°C) [51] and then to many more substances. In fact, zone refining can be applied to the purification of any material that melts congruently. The technique has also been used to control the discontinuities in impurity in p–n or n–p junction.

Actually, zone refining is one of the several techniques grouped under the general class "zone-melting." Other techniques in this class are zone freezing, zone leveling, zone separation, float zone technique, traveling solvent method, and so on. Although developed by Pfann for the purpose of purification, it could be easily adapted to growing crystals and hence has now become an established method of growing high-purity single crystals.

Normal and Equilibrium Freezing

To understand the process of purification, we need to introduce a term known as segregation or distribution coefficient k. Referring to a simple binary phase diagram, as shown in Figure 8.25, let us focus our attention to the regions close to the points T_A or T_B, where one of the components is major in which the other exists as an impurity/solute. The expanded version of the regions around the point T_A and T_B are shown in Figure 8.26a and b, respectively. We see from Figure 8.26 that at any particular temperature the concentration of the impurity in the major phase depends on whether the major phase is in the solid or in the liquid form. In fact, the concentration of the impurity in the solid phase is either less (Figure 8.26a) or more (Figure 8.26b) than that in the liquid phase. Now the distribution coefficient is defined as the ratio between the concentration of the impurity in the solid phase and that in the liquid phase. Since we

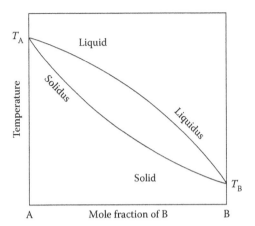

FIGURE 8.25
Binary phase diagram of a solid solution.

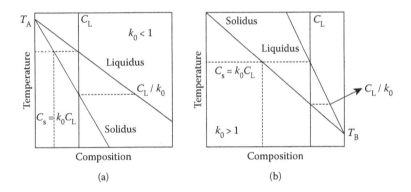

FIGURE 8.26
Expanded version around points (a) T_A and (b) T_B in Figure 8.25.

are defining this term with respect to the equilibrium phase diagram, it is called the equilibrium distribution coefficient k_0. That is,

$$k_0 = \frac{C_S}{C_L} \tag{8.2}$$

Now, in most cases, this quantity will either be greater or less than 1. It will be equal to 1 only when the liquidus and solidus curves collapse on each other, which will be a rare case.

Let us consider a complete freezing of a melt that is in the form of a long cylinder freezing from one end to the other end, as shown schematically in Figure 8.27a. This operation is referred to as normal freezing. In this case,

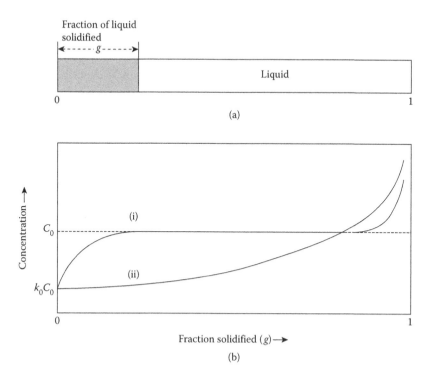

FIGURE 8.27
(a) Solidification by normal freezing, and (b) impurity distribution as a function of fraction solidified, (i) complete mixing in liquid and (ii) mixing by diffusion only.

the distribution of solute (impurity) between solid and liquid during freezing and throughout the solid after freezing will depend on not only k_0 but also the conditions of freezing. The two most important variables are the solidification rate (i.e., the rate of advancement of the solid–liquid interface) and the degree of the mixing in the liquid. Depending on the strength of these variables, the solute distribution process may follow different courses. One extreme case is the equilibrium freezing. This occurs if the gradients in concentrations and temperature in the system are negligibly small, which implies that the freezing is extremely slow, slow enough to permit diffusion process in the solid and liquid to erase any concentration gradients. As the system undergoes equilibrium freezing, the concentration of solute in the solid is always k_0 times that in the liquid. Since the freezing action rejects solute into the liquid, its concentration in the liquid and the freezing solid continually rises. This is shown in Figure 8.27b as a continuously rising curve. However, equilibrium freezing is very rarely achieved in practice because diffusion rates in solids are prohibitively slow. If no diffusion occurs in the solid, then segregation will occur during normal freezing. The conditions of no diffusion in the solid and uniform concentrations in the liquid lead to maximum segregation during normal freezing (curve ii in Figure 8.27b). These conditions can be approached by selecting a freezing rate that is large compared to the diffusion rate of the solute in solid and yet small compared to that in liquid.

Impurity distribution caused by normal freezing can be described analytically by an expression

$$C = kC_0 (1 - g)^{k-1} \tag{8.3}$$

in which C is the concentration of the impurity in solid at the point when the fraction g of the original liquid has frozen. C_0 is the mean initial concentration.

Assuming a constant value of k for all concentrations of impurity will lead to erroneous results because of the difference in the mutual inclination of the solidus and liquidus curves for different values of concentrations. Therefore, for each value of k it is also necessary to specify the appropriate concentration since the solidus and liquidus curves can generally be represented by straight lines only over a very narrow range of solute concentration.

The assumption that equilibrium exists between concentrations of impurity in liquid and solid is incorrect because of lack of complete mixing in the melt. Since the rate of impurity rejection at the advancing interface is faster than the impurity diffusion into the main body of the liquid, a steady-state stagnant boundary layer of certain thickness δ (see Figure 8.28) will have to be assumed to arrive at an effective distribution coefficient. Burton et al. [52] gave the following equation for the effective distribution coefficient, taking the boundary layer into account.

$$k_{\text{eff}} = \frac{k_0}{k_0 + (1 - k_0)\exp\left(-f\delta / D\right)} \tag{8.4}$$

where k_0 is the equilibrium distribution coefficient, f is the rate of freezing, D is diffusion coefficient of solute in the liquid, and δ is the thickness of the

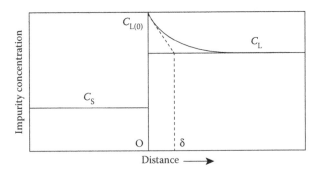

FIGURE 8.28
Concentration of impurity at the melt solid interface creating a boundary layer for k < 1 (rate of freezing >> rate of diffusion).

boundary layer. This equation predicts that $k_{eff} \rightarrow k_0$ as $f \rightarrow 0$ and $k_{eff} \rightarrow 1$ as $f \rightarrow \infty$. Thus, as the rate of freezing increases, segregation diminishes.

Principle of Zone Refining

Segregation due to Single-Zone Pass

The principle of zone refining can be understood easily with the help of Figure 8.29a. Here, the charge is comprised of a major phase in which the impurity phase is distributed homogeneously with the initial concentration C_0 and is in the form of a cylindrical rod. Let the distribution coefficient k for the impurity in the major phase be less than 1. Let a molten zone of length l be traversing the charge slowly. The impurity distribution at the end of the single-zone pass is represented by Figure 8.29b. It has three regions: the initial purified region, the level region, and the final impure region. The reason for its occurrence is straightforward. As the molten zone advances from one end, the first layer that solidifies at $x = 0$ will have its impurity concentration as kC_0 throwing a certain amount of impurity into the melt, thereby enriching the zone. This enrichment continues until the impurity concentration in the zone reaches a value C_0/k. From then onward the input and output of impurity concentration will be the same until the zone reaches the end of the rod. Further forward movement of the melt zone decreases its length, resulting in the sharp rise in the impurity concentration.

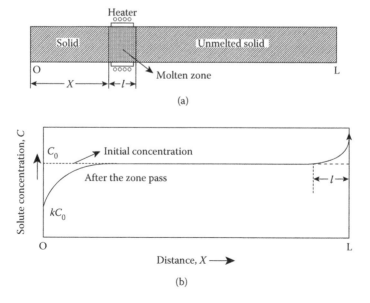

FIGURE 8.29
(a) Molten zone passing through a cylindrical charge rod and (b) approximate impurity concentration after a single-zone pass.

The equation for the process mentioned earlier as a function of x is given by Pfann [50]

$$(C_x/C_0) = 1 - (1-k)e^{-kx/l} \qquad (8.5)$$

where C_x is the concentration of impurity in the solid at a distance x from the starting end. This equation is valid at all the distances except for the last zone length, where the solidification occurs by normal freezing. As can be inferred from Equation 8.5, the refining will be more effective as k deviates more from 1.

In the earlier discussion, we have assumed k to be less than 1. If $k > 1$, the impurity will accumulate at the starting end of the rod and the far end will have a lower concentration of impurities. Thus, the refining still occurs.

Segregation due to Multiple-Zone Passes

The real merit of zone refining is realized when the process is repeated successively. One can qualitatively visualize what would happen when the molten zone is passed for the second time through the ingot. With the initial impurity concentration of the first layer of the ingot being $k_0 C$, second solidification, in principle, will bring down the impurity concentration of the first solidifying layer to $k^2 C_0$. Consequently, the initial purified region gets longer for the second zone pass, and the corresponding amount of impurity will be transported to the other end of the ingot. As the number of zone passes increases, the initial end would get more and more purified and the impurity will accumulate at the far end. Figure 8.30 shows schematically the impurity profile as a function of distance with increasing number of zone passes.

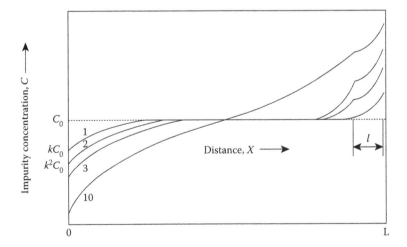

FIGURE 8.30
Concentration of the solute along a zone-refined bar for several zone passes.

Instead of repeating the zone passes with just one heater, one can use a set of closely spaced heaters and traverse the entire set along the ingot, thereby achieving multiple-zone passes in a single operation.

Zone Leveling

In zone leveling, the impurity level is homogenized by moving the zone back and forth along a straight rod. Figure 8.31 schematically represents the impurity profile for one forward and reverse zone pass for $k < 1$. As the number of forward and reverse zone passes increases, the impurity concentration becomes more uniform. Zone leveling can also be carried out by moving a zone around a ring-shaped charge over and over. A continuous cyclic process is used usually when nothing is known about the nature of impurity.

Zone Freezing

Zone freezing is really the reverse of zone-melting. Here, a narrow crystalline zone is passed through the melt by traversing a cooling coil (Figure 8.32). Substances that lower the melting point will be in the melt. For $k < 1$, the impurity concentration will be less in the crystalline zone; hence, the zone practically pushes the impurity to the melt in front of it. Hence, impurities are swept away to one end. Obviously, the impurity concentration will be the lowest in the melt behind the crystalline zone. However, as the crystalline zone advances, the impurity concentration in it slowly rises and hence that of the melt behind it also increases. Since the redistribution of the impurities is faster in the liquid, the impurity concentration will eventually be uniform in the melt behind the crystalline zone. Hence, in a way zone, freezing has the combined effect of zone refining and zone leveling.

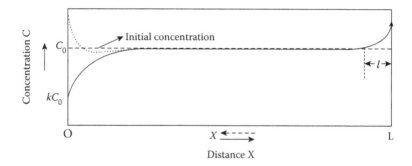

FIGURE 8.31
Distribution of impurity after one forward (continuous line) and reverse (dotted line) zone pass, $k < 1$.

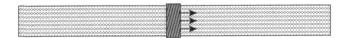

FIGURE 8.32
Zone freezing: a narrow zone of crystalline solid is moving through the long molten sample.

Zone Refining as a Technique for Crystal Growth

Although zone-melting was developed more as a purification technique rather than a crystal growth technique, the two are often performed together when one wants crystals of very high purity. As the number of zone pass increases, the grain size increases and finally results in a single crystal. Often dislocation-free crystals are grown this way. It is worth remembering that Kapitza [48] used the zone-refining method to grow high-purity bismuth single crystals. A few years later Andrade and Roscoe [53] also used the traveling zone method to grow single crystals of cadmium and lead.

In the case of single crystal preparation, a seed crystal is normally used. Even when a single crystal seed is not available, crystal can be grown through grain selection in which case the initial portion of the zone-refined ingot will be polycrystalline. As the molten zone advances, grain selection occurs in the solidified ingot and eventually a single crystal grows.

The most commonly used heating is by either direct or indirect (with the help of a susceptor) radio frequency heating. Other methods include electron beam melting, electric arc, and plasma heating, as well as thermal and light radiation. Radio frequency coils should be of short length. Often single coil would be sufficient. With nonconducting materials, preheating or a separate susceptor may have to be employed. For oxides where conductivity is low, often a strip heater is used. In this technique, a precious metal strip usually of platinum is resistively heated and is passed through the molten zone. The strip has several holes in it, through which the molten oxides can flow when the zone travels through the charge.

Horizontal Configuration

The zone-refining technique is used both in horizontal and vertical configurations. Figure 8.33a shows a schematic of a horizontal zone-melting technique. In this configuration, a boat is used to hold the charge. The boat with the charge and a seed at one end are placed inside a clear fused silica tube. A few turns of an induction coil that is wound over this tube form the heater. Initially, a narrow molten zone is formed at the seed end so that the seed in contact with the charge partially melts to form a homogeneous solid–liquid interface. The boat is subsequently translated at a slow rate so that the molten zone traverses the entire charge, leaving behind a crystalline ingot. Figure 8.33b shows a homemade horizontal zone-refining unit

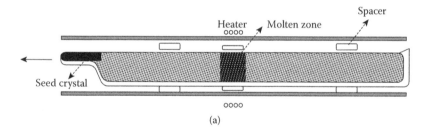

(a)

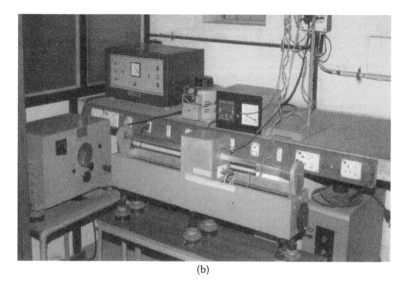

(b)

FIGURE 8.33
(a) Schematic of the horizontal configuration and (b) a homemade horizontal zone-refining unit.

with resistive heating. In this unit, the heater is traversed and the charge is held stationary inside a transparent quartz tube.

Material Transport

When a molten zone traverses the charge kept horizontally in an open boat, the resulting ingot is often found tapered and the extent of tapering is found to increase with the number of zone passes. Such material transport during zone-melting is due to the density difference between the melt and the solid. The direction of the material transport depends on the sign of the density change. If the material contracts on melting, the material transport is in the direction of the zone pass, as shown in Figure 8.34a. If it expands, the transport is opposite to the direction of the zone pass (Figure 8.34b). The magnitude of tapering depends on the magnitude of the density change.

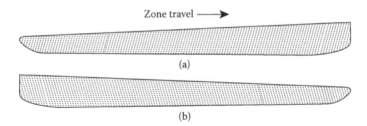

FIGURE 8.34
Material transport due to zone passes: (a) density of solid greater than that of liquid and (b) density of liquid greater than that of solid.

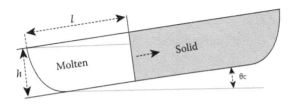

FIGURE 8.35
Boat inclined at a critical angle for zero material transport.

The material transport can be prevented by suitably tilting the boat with respect to the horizontal by a proper angle (Figure 8.35). It can be shown that the tilting angle θ is related to the ratio of the density of the solid to that of the melt, α, by the following expression [50]:

$$\theta = \tan^{-1} 2h(1- \alpha)/l \tag{8.6}$$

where h and l are as depicted in Figure 8.35. In deriving this equation, the effect of surface tension, wetting, stirring, and so on is not taken into account. A more detailed analysis taking into account surface tension and wetting has been made by Schildknecht [54].

As we have seen, for zone refining in horizontal configuration, crucible is essential. This can lead to reaction with melt, particularly at high temperatures, or the grown crystal might stick to the boat. The vertical configuration, commonly called the float zone technique, which is a crucible-free method, overcomes these problems.

Vertical Configuration: Floating Zone Technique

The floating zone technique was essentially developed to circumvent the problem of contamination of crucible material into the growing crystal prevalent in the horizontal version. In this technique, the material is in the form of a free-standing polycrystalline rod clamped vertically at its two

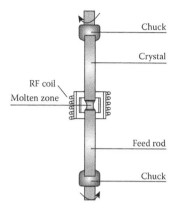

FIGURE 8.36
Schematic of vertical zone-melting (float zone-melting).

ends (Figure 8.36). A small zone in this rod is melted by a suitable heater. The melt is suspended like a drop between the two parts of the rod; hence, the melt is in contact with its own solid, and the technique does away with the container. The crucible-free situation has an additional purifying effect because the impurities can evaporate from the melt surface without any hindrance. Since the melt is essentially floating, the technique is called the floating zone technique. Here, the surface tension holds the molten zone of liquid in between the two parts of the sample. The molten zone is moved through the entire rod either by moving the heater or by the rod itself. Single crystals can be prepared either by spontaneous nucleation or by using a single crystalline seed at the end of the feed rod. In the former case, a certain length of the zone-refined rod would be polycrystalline. As the growth advances, the grain selection takes place and eventually single grain will succeed. A stirring effect is caused by rotation of the two ends of the sample in the opposite direction.

Zone Stability

The shape and stability of the molten zone are crucial to the floating zone technique. As mentioned earlier, the molten zone is held against the gravity essentially by surface tension. Hence, materials with low liquid density and high surface tension are preferred. In principle, if the length of the molten zone exceeds its circumference, the zone will become unstable. This stability criterion is, however, valid only when the weight of the melt is neglected, which would mean that the diameter of the zone is very small and the growth is occurring in zero gravity conditions. In the presence of gravity, however, the molten zone develops meniscus, which may be either convex or concave. The conditions for zone stability have been well studied and documented by Heywang and Ziegler [55,56]. Heywang has empirically arrived at the following conditions for zone stability for thin and thick rod situations. According to him, the maximum stable zone length for thin rod is

$$l_m = 2\sqrt{3r} \qquad (8.7)$$

where r is the radius of the rod.

For thick rod, the stability condition is

$$l_m = 2.8\left(\frac{\gamma}{\rho g}\right)^{1/2} \qquad (8.8)$$

where γ is the surface tension, ρ is the density of the melt, and g is the acceleration due to gravity.

Example

In this section, specific details of the preparation of single crystals of DyMnO$_3$ using the optical floating zone furnace are presented [57]. DyMnO$_3$ is an interesting multiferroic crystal that can be synthesized in both orthorhombic and hexagonal polymorphs.

Material Synthesis and Preparation of Ingots

The first step in a growth experiment using the floating zone furnace is the solid-state synthesis of the compound with the desired composition. For example, in the preparation of DyMnO$_3$, the starting materials are taken as per the following chemical reaction:

$$2Dy_2O_3 + 4MnO_2 \rightarrow 4DyMnO_3 + O_2\uparrow$$

Before weighing, Dy$_2$O$_3$ is preheated to remove moisture. The stoichiometric mixture of Dy$_2$O$_3$ and MnO$_2$ is ground in a ball mill or a mortar for 1–2 h and sintered at 1200°C with repeated intermediate grinding. This process of grinding and sintering is repeated several times to ensure a homogeneous chemical phase. The phase purity of the synthesized material is verified by taking powder x-ray diffraction patterns.

The second step is the fabrication of ceramic ingots. The synthesized powder is carefully filled in a rubber tube and placed inside stainless steel split tubes and fastened tight with cellophane tape. Rubber tubes and the stainless steel casings used for preparing the ceramic ingots are shown in Figure 8.37. This arrangement is then inserted vertically into a hydrostatic press and subjected to ~70 MPa pressure. The pressed ingots are sintered again at high temperature (1200°C) by placing them in ceramic boats or on platinum foils. Inhomogeneous temperature profiles in the furnace can

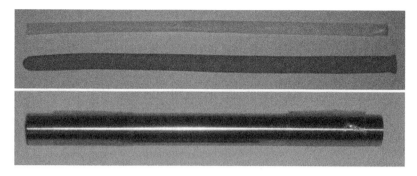

FIGURE 8.37
Rubber tubes and the stainless steel casing that are used in ingot preparation.

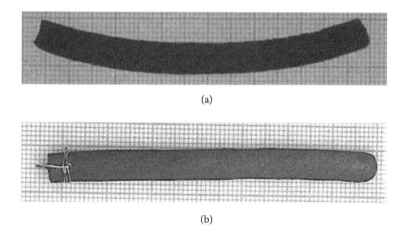

(a)

(b)

FIGURE 8.38
(a) A bent feed rod, disfigured due to inhomogeneous temperature profile of the sintering furnace, and (b) a useable straight feed rod with platinum wire attached.

lead to bending of the ingots and render them useless because they cannot be aligned vertically inside the growth chamber. A bent ingot is shown in Figure 8.38 along with a useable one.

Typical Growth Run

For growing $DyMnO_3$ crystals, the optical furnace FZ–T–10000–H–VI–VP from Crystal Systems Inc., Japan, is used and the same is shown in Figure 8.39a along with its schematic (Figure 8.39b). It is equipped with four halogen lamps that can deliver a total power of 6 kW. The lamps are located at one of

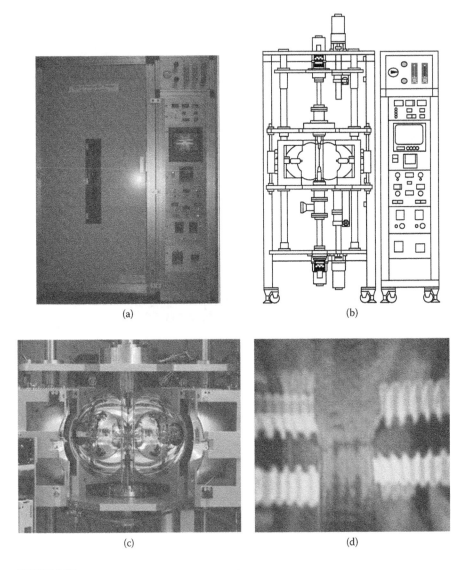

FIGURE 8.39
(a) Optical floating zone furnace FZ-T-10000-H-VI-VP (Crystal System Inc., Japan); (b) schematic of the furnace; (c) internal view of the four-mirror furnace; and (d) melt zone as viewed from the monitor. (Courtesy of Crystal System Inc., Japan.)

the foci of the hemi-ellipsoidal mirrors. The other focus is the point where the radiation from the four mirrors converges. Figure 8.39c shows the internal view of the four-mirror furnace. The four-mirror arrangement provides a better homogeneous isothermal region than that provided by conventional single- or two-mirror furnaces.

To begin, one of the ceramic ingots is fastened to a ceramic holder using nichrome wire and is fixed to the bottom rotating shaft of the optical floating zone furnace. This acts as the seed rod. When an oriented single crystal is available, it can be used as a seed. The purpose of the seed rod is to provide a pedestal support to the molten zone. Another ceramic ingot is suspended from a platinum hook attached to the top rotating shaft of the optical floating zone furnace. This acts as the feed rod. A platinum wire inserted through the hole drilled at one end of the feed rod helps to align it such that it rotates about its own axis. The growth chamber is enclosed by a quartz tube, which helps to employ different gases as ambiance and also protects the mirrors from accidents due to melt spilling.

Initially, the feed and the seed rods are aligned vertically inside the quartz tube with their tips separated from each other by a few millimeters. Then, the growth chamber is sealed and the desired ambiance (air, nitrogen, oxygen, or argon) is introduced. The rods are set to rotate in opposite direction, typically at 20–30 rpm. Heating is commenced by switching the halogen lamps on. A Eurotherm temperature controller regulates the percentage power supplied to the lamps rather than controlling the temperature directly. As the heating progresses, the tips of the rods are brought to the hot zone and melted. Now, the temperature of the zone is manually raised through the controller in small steps of 0.1%–0.5% in power. Also, the rotation rate of the feed and seed rods is increased appropriately to facilitate uniform melting. This is continued until the tip of the feed rod forms a convex shape. At this stage, the rods are brought closer to each other and contact is established. The melt volume is adjusted such that the length of the molten zone is approximately equal to or slightly less than the diameter of the rods. As mentioned earlier, this is crucial for zone stability. After forming the molten zone, a few minutes is spent to stabilize it by fine tuning parameters like rotation rate, power, and so on. The growth is then commenced by starting the translation of mirror stage. Figure 8.39d shows the image of molten zone captured by the CCD camera. Typical growth rates of oxides are 5–8 mm/h. Hence, to obtain a single crystal of appreciable dimensions, the experiment has to be continued for 20 h or so. After the desired length of the crystal is attained, the experiment is terminated. This is performed in gradual steps of reduction of power accompanied by pulling apart the seed and feed rods to reduce the melt volume. This procedure is carefully carried out to avoid spilling of melt. After this, the cooling protocol is programmed in the temperature controller. Typically, slow cooling rates in the range of 5%–15% power/min are used. Usually, several growth runs have to be performed to optimize various growth parameters to obtain a single crystal. It is known that $DyMnO_3$ can be crystallized in two crystallographic variants (i.e., hexagonal and orthorhombic) depending on the ambient gas used during growth [58,59]. The hexagonal and the orthorhombic varieties are obtained by performing the growth in argon and air, respectively. A photograph of the $DyMnO_3$ crystal grown in argon atmosphere is shown in Figure 8.40 [57].

FIGURE 8.40
Single crystal of $DyMnO_3$ along with the seed rod. (From Harikrishnan, S., Phase transition and magnetic order in multiferroic and ferromagnetic rare earth manganites, PhD dissertation, Indian Institute of Science, Bangalore, India, 2009.)

Flame Fusion Technique

The flame fusion technique is a well-established technique for the growth of refractory oxide crystals. The process was developed by the French chemist Verneuil in 1902 [60] and became the first commercially successful method of manufacturing synthetic gemstones. In this technique, a flame produced by an oxy-hydrogen torch melts the powder, which rains down into a shallow pool of liquid on the top of a seed crystal; this is slowly lowered down into an after heater to produce a cylindrical crystal. The technique has the advantage of being crucible-less. An extended review article on the process has been written by Falckenberg [61] and can be found in the book edited by C. H. L. Goodman.

Figure 8.41 shows a schematic of a Verneuil system. In a typical growth run, finely powdered starting material is placed in a container within a Verneuil system, with a sieve at the bottom through which the powder can rain down when the container is tapped or vibrated. While the powder is being released, oxygen is supplied into the furnace and travels with the powder down a narrow tube. This tube is surrounded by a larger tube, into which hydrogen is supplied. At the point where the narrow tube opens into the larger one, a flame is formed due to combustion with a temperature of more than 2000°C at its core. As the powder passes through the flame, it melts into small droplets, which fall onto a pedestal placed below, forming a shallow pool of melt. As more droplets fall onto it, the pedestal is slowly moved downward, allowing the lower portion of the liquid to crystallize, while its cap always remains liquid. The crystal is formed in the shape of a tapered cylinder, with its diameter broadening slowly from the base and eventually remaining more or less constant. With a constant supply of powder and withdrawal of the pedestal, very long cylindrical boules can be obtained. During the growth, the amount

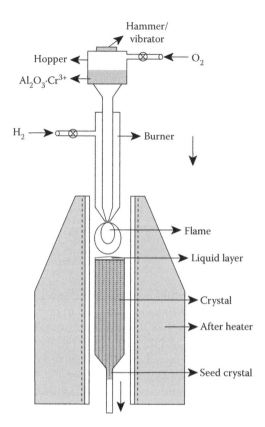

FIGURE 8.41
Schematic of the flame fusion technique.

of material equal to that added to the melt crystallizes on the underside of the shallow melt, thereby maintaining a constant melt thickness. For this to happen, the crystal lowering rate and the powder feeding rate should be synchronized. When the desired crystal length is reached, the process is terminated by stopping the powder flow and lowering the crystal, and the flame is extinguished. Usually the flame and the crystal are coaxial and to ensure thermal symmetry, seed crystals are rotated at 5–50 rpm. Rotation can be done at right angles to the flow to get discs. Once removed from the furnace and allowed to cool, the boule is often split along its vertical axis to relieve internal stresses. One such split ruby crystal is shown in Figure 8.42.

Crystals produced by the Verneuil technique are found to be physically and chemically equivalent to their naturally occurring counterparts. In fact, it is often difficult to distinguish between the two. As far as the perfection is concerned, the crystals grown by this technique are inferior to those grown by others such as the Czochralski technique with typical dislocation density of 10^5 cm^{-2}.

FIGURE 8.42
A ruby crystal split along the boule axis.

Technical Considerations

While discussing the process in detail, Verneuil [62] specified several conditions crucial for good growth. These were with regard to optimum flame temperature necessary for fusion, position of the melt in the oxy-hydrogen flame (to be maintained at the same level), and the size of the point of contact between the melt and support pedestal. These and the others evolved over the years, leading to the following technical considerations: (a) seed preparation, (b) burner design, (c) powder preparation and feeding, and (d) growth chamber. These are discussed in the following sections.

Seed Preparation

Since many of the crystals grown by the Verneuil technique have their natural counterparts, seeds may be prepared from them. In the absence of such an option, they may be obtained by a spontaneous nucleation process. When the melt solidifies over the pedestal, it is usually polycrystalline, but grain selection leading to a single grain can be achieved by narrowing the boule cross-section; when a single crystallite dominates the interface, the cross-section can be gradually increased. Readers might recollect that this procedure is similar to the necking down employed in the Czochralski technique. Such crystals can be used for successive growth as seeds after properly orienting them.

Burner Design

There have been many improvements in burner design. In almost all the cases, oxy-hydrogen flame is the preferred one, although gas mixtures such as O_2–CH_4, O_2–CO, and O_2–acetylene have been occasionally used. Since oxygen and hydrogen are not to be mixed before their ignition, the gases are

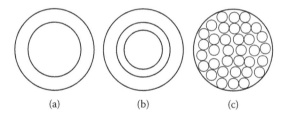

FIGURE 8.43
Cross-sectional view of a (a) two-cone burner, (b) tricone burner, and (c) multitube burner.

always led through separate tubes until the ignition point. Burners usually consist of concentric stainless steel tubes. In the simplest of the cases, there would be two concentric tubes (Figure 8.43a), with the inner one carrying the powder and oxygen while the outer tube carries hydrogen. The other modification is the tricone burner due to Merker [63]. In this design, three concentric tubes are used (Figure 8.43b). The inner tube carries hydrogen while the outer two tubes carry oxygen. Consequently, the flame will be oxygen-rich and hence forms an oxidizing atmosphere. Also, this produces a pointed flame that is useful to initiate growth without seed where a small quantity of the material is to be initially melted. However, temperature in this case is less than the normal H_2–O_2 flame. One other modification is a multitube burner, which has a number of small tubes inside a large tube (Figure 8.43c). Small tubes carry O_2 and powder, and the large tube carries H_2 gas. This produces a large flame front with fairly uniform flame temperature, which can be used to grow large-diameter crystals.

Other methods of heating such as the plasma torch have also been used in the Verneuil process.

Powder Preparation and Feeding

One of the most crucial factors in successfully growing an artificial gemstone is the purity of the starting material. Starting material with at least 99.9995% purity should be used. For example, starting material for growing sapphire or ruby is alumina. The presence of a small quantity of sodium impurities in alumina makes the crystal opaque and hence should be avoided. Depending on the desired color of the crystal, small quantities of various oxides are added as dopants, such as chromium oxide for a red ruby or ferric oxide and titanium oxide for a blue sapphire. However, one should mix the dopant as uniformly as possible because the homogeneity of the dopant distribution depends on it. The charge should be in the form of powder. During the fusion process since there is a reduction in volume due to phase change (by a factor of 20 in the case of Al_2O_3), course powder leads to void formation. The formation of voids decreases as the particle size decreases. Thus, the powder should be fine but not so fine that it does not flow freely. Those trapped between 120 and 300 mesh (50–125 µm) are typical. If one of the constituents of powder is more volatile, it should be taken in excess. For example, while growing $SrTiO_3$, 3% excess of $SrCO_3$ is normally taken. Powder must be dry. To ensure this, the powder chamber is heated to a temperature of about 100°C.

Uniform feeding of the powder is a crucial aspect of the Verneuil technique. The short-term constancy of the powder flow should be as high as possible. Consequently, many feeding mechanisms have been developed, which range from uniform tapping to two-stage vibratory feeders with combined horizontal and vertical motion. The tapping mechanism, which in fact was originated by Verneuil himself, is still widely used in industry on account of its simplicity and reliability. In this mechanism, the powder container is simply tapped with a hammer at regular intervals. The throughput essentially depends on the force with which the hammer taps the container and the frequency of tapping and decreases with the sieve aperture independently of the hammer weight. The return motion of the sieve base following each hammer blow results in a suction effect causing a momentary stoppage of the powder flow. Consequently, short-term constancy of powder feeding is poor. To improve the constancy in powder feeding, several two-stage devices have been developed, details of which can be found in the article by Falckenberg [61].

Growth Chamber

Growth chambers of various designs can be found in the literature; they were developed to grow certain specific crystal or sometimes group of crystals. It may be noted that the temperature of operation in the Verneuil process is 2000°C or above. Hence, only those refractory materials that can withstand such high temperatures should be used in chamber fabrication. In most cases, tubular furnaces are preferred. If a viewing port is provided in the chamber, the growing crystal can be viewed to monitor the process.

Growth chambers are often augmented with after heaters. This is primarily to prevent the crystal from cracking due to the rapid cooling process that the crystal undergoes when being withdrawn from the flame during growth. The presence of an after heater reduces the thermal gradient significantly, thereby preventing a thermal quench. Such after heaters can be of resistive type such as platinum wire wound on the lower part of the tubular growth chamber [64]. Subsidiary heating can also be by gas. For example, to grow $MgAl_2O_4$, Mitchell [65] has used an additional gas heater of oxygen-propane gas flame.

Typical Crystals Grown

The Verneuil growth technique is primarily used to produce ruby and sapphire, varieties of corundum, and rutile and strontium titanate, but it is

TABLE 8.2

Representative Crystals Grown by the Verneuil
Technique

Material	Melting Point (°C)	Reference
TiO_2	1830	[66]
$Al_2O_3 \cdot Cr^{3+}$	2040	[67]
$SrTiO_3$	2080	[68]
$Mg\ Al_2O_4$	2130	[69]
Y_2O_3	2400	[70]
ZrO_2	2700	[71]

not limited to these. Some crystals grown early by this technique are listed below along with their melting points and references (Table 8.2).

In summary, the Verneuil method has proved to be a successful and cheap technique for growing a variety of refractory crystals. While the quality of the crystals produced may not be as good as the ones grown by other melt techniques, it has been realized in recent years that this is not inherent to the technique itself. Further research into its improvement in terms of design and control parameters should improve the quality of the grown crystals by this method.

Crystal Growth by the Arc Fusion Technique

This method, although crude, is useful to produce low-cost crystals of high melting point materials. Since there is not much control in this technique, the resulting crystals are poor in quality. The schematic of the arc fusion technique is shown in Figure 8.44. The method basically involves striking an arc in a large quantity of charge in the form of compacted powder taken in a water-cooled steel closure or even in a firebrick kiln [72,73]. Usually graphite electrodes are used. Once a pool of molten material is formed due to arcing, the power is slowly reduced to stop the arc, and the melt is left to solidify. The solidified mass, which contains many crystals, is broken to harvest the crystals. Crystals weighing up to half a kilogram have been grown. Since the melt is in contact with its own solid, there is no contamination due to crucible material. The shell left by taking out the crystals may be reused by refilling it with fresh material, thereby economizing the process. Introduction of certain gases sometimes improves the quality of the grown crystals. For example, it is found that by introducing argon through the hollow electrode, one could obtain much clearer MgO crystals. Improvement in terms of number of electrodes used has been reported in the literature [74]. Other crystals grown by the arc fusion method are CaO, SrO, BaO, and ZrO_2.

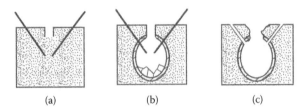

(a)   (b)   (c)

FIGURE 8.44
Schematic of the arc fusion technique: (a) the enclosure with the charge loaded, (b) after the solidification of the melt, and (c) after the removal of the crystals.

Growth by the Skull-Melting Technique

The skull-melting technique, also called the cold crucible technique, is particularly useful to grow refractory oxides. The technique allows the melt to be held at very high temperatures (3000°C or even above). Actually, skull-melting as a crystal growth process must have evolved from growth of crystals in water-cooled crucibles using the Czochralski method. A review by Osiko et al. [75] in the *Springer Handbook of Crystal Growth* discusses the skull-melting technique in detail with special reference to the growth of cubic zirconia. There is also a book by the same group titled *Cubic Zirconia and Skull Melting*, published by Cambridge International Science Publishing Ltd [76].

Keeping the melt in a solid shell (skull) with a chemical composition identical to that of melt and using a contact-free method of heating the material are the two special features of this technique. Since melting is carried out through radio frequency heating using an induction furnace, superficially it looks as if the method is useful only to grow such materials which are conducting. However, most materials become conducting when hot and hence couple to the radio frequency source and eventually melt. Normally, metal of the same oxide that is to be melted to obtain crystals is used for initial coupling. Alternately, conducting cylinder-like graphite can be inserted into the insulating charge, which can be removed once the charge becomes hot and starts melting.

The cold crucible usually consists of interconnected hollow sections with gaps, fabricated in the form of a cylindrical crucible (Figure 8.45). The crucible is kept cold by the flowing water. The leakage of the molten charge is prevented by the surface tension and wetting property of the melt as well as the size of the gaps. Materials used for water-cooled crucible are copper, brass, chromium plated brass, nickel, platinum-plated nickel, stainless steel, rhodium-plated stainless steel, and so on. In a typical experiment, the charge is loaded into the crucible. The charge in the case of zirconia is a mixture of zirconium oxide, a stabilizer such as

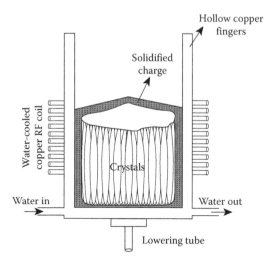

FIGURE 8.45
Schematic of a cold crucible.

TABLE 8.3

Certain Refractory Crystals Grown
by the Skull-Melting Techniques

Material	Melting Point (°C)
MgO	2852
ZrO$_2$	2750
CaO	2614
SrO	2430
BaO	1918

yttrium oxide or calcium oxide to stabilize the cubic phase of ZrO$_2$, and a small amount of zirconium metal itself. The crucible is wrapped with a radio frequency coil, which heats the charge. Once the charge is melted and homogenized, during which period Zr metal also gets oxidized to become ZrO$_2$, crystallization is initiated either by rapid solidification or by directional solidification as in Bridgman, the former method being used for mass production. Even the crystal pulling method has been employed. Table 8.3 lists some refractory crystals grown by the skull-melting technique.

The advantage of the cold crucible technique over the arc fusion method discussed in the previous section is the possibility of better temperature control through power. In the radio frequency system, the power can be controlled to one part in several thousands fairly easily, while the power in the arc is difficult to control to better than few parts per hundred.

References

1. Bridgman, P. W. 1925. Certain physical properties of single crystals of Tungsten, Antimony, Bismuth, Tellurium, Cadmium, Zinc, and Tin. *Proceedings of the American Academy of Arts and Sciences* 60: 305–383.

2. Stockbarger, D. C. 1936. The production of large single crystals of lithium fluoride. *Review of Science Instruments* 7: 133–136.

3. Stockbarger, D. C. 1949. The production of large artificial fluorite crystals. *Discussions of the Faraday Society* 5: 294–299.

4. Brice, J. C. 1986. *Crystal growth processes.* Glasgow, UK: Blackie.

5. Hurle, D. T. J., ed. 1994. *Hand book of crystal growth.* Amsterdam: North Holland.

6. Dutta, P. S., K. S. Sangunni, H. L. Bhat, and V. Kumar. 1994. Growth of gallium antimonide by vertical Bridgman technique with planar crystal-melt interface. *Journal of Crystal Growth* 141: 44–50.

7. Udayashankar, N. K., K. Gopalakrishna Naik, and H. L. Bhat. 1999. The influence of temperature gradient and lowering speed on the melt–solid interface shape of $Ga_xIn_{1-x}Sb$ alloy crystals grown by vertical Bridgman technique. *Journal of Crystal Growth* 203: 333–339.

8. Rout, R. K., M. Wolf, and R. S. Feigelson. 1984. Interface studies during vertical Bridgman CdTe crystal growth. *Journal of Crystal Growth* 70: 379–385.

9. Crouch, R. K., W. J. Debman, and R. Ryan. 1982. Vacuum tight quartz ampoule for bridgman growth of crystals with interface demarcation. *Journal of Crystal Growth* 56: 215–216.

10. Dutta, P. S., K. S. Sangunni, H. L. Bhat, and V. Kumar. 1994. Experimental determination of melt-solid interface shapes and actual growth rates of gallium antimonide grown by vertical Bridgman method. *Journal of Crystal Growth* 141: 476–478.

11. Campbell, T. A., and J. N. Koster. 1997. In situ visualization of constitutional supercooling within a Bridgman-Stockbarger system. *Journal of Crystal Growth* 171: 1–11.

12. Venkataraghavan, R., K. S. R. K. Rao, and H. L. Bhat. 1997. The effect of growth parameters on the position of the melt—Solid interface in Bridgman growth of indium antimonide. *Journal of Physics D: Applied Physics* 30: L61–L63.

13. Dutta, P. S., H. L. Bhat, and V. Kumar. 1997. The physics and technology of gallium antimonide: An emerging optoelectronic material. *Journal of Applied Physics* 81: 5821–5870.

14. Dixit, V. K., and H. L. Bhat. 2010. Growth and characterization of antimony-based narrow-bandgap III-V semiconductor crystals for infrared detector applications. In *Springer handbook of crystal growth*, edited by G. Dhanaraj et al. Heidelberg: Springer, pp. 327–366.

15. Udayashankar, N. K., and H. L. Bhat. 2001. Growth and characterization of indium antimonide and gallium antimonide crystals. *Bulletin of Materials Sciences* 24(5): 445–453.

16. Czochralski, J. 1917. Ein neu es Verfahren zur Messung der Kristallisationsgeschwindigkeit der Metalle [A new method for the measurement of the crystallization rate of metals]. *Zeitschrift fur Physikalische Chemie* 92: 219–221.

17. Kyropoulos, S. 1926. Ein Verfahren zur Herstellung grosser Kristalle. *Zeitschrift fur Anorganische und Allgemeine Chemie* 154: 308–313.
18. Kyropoulos, S. 1930. Dielektrizitätskonstanten regulärer Kristalle. *Zeitschrift fur Physik* 63: 849–854.
19. Bardsley, W., G. W. Green, C. H. Holliday, and D. T. J. Hurle. 1972. Automatic control of Czochralski crystal growth. *Journal of Crystal Growth* 16: 277–279.
20. Nassau, K., and L. G. Van Uitert. 1960. Preparation of large calcium-tungstate crystals containing paramagnetic ions for maser applications. *Journal of Applied Physics* 31: 1508–1508.
21. Abrahams, S. C., ed. 1989. Properties of Lithium Niobate. *Electronic materials information service (EMIS) Data reviews.* Series No.5. INSPEC.
22. Prokhorov, A. M., and Y. S. Kuzminov. 1990. *Physics and chemistry of crystalline lithium niobate.* Bristol: Adam Hilger.
23. Yan, W., Y. Kong, L. Shi, L. Sun, H. Liu, X. Li, D. Zhao, et al. 2006. The relationship between the switching field and the intrinsic defects in near-stoichiometric lithium niobate crystals. *Journal of Physics D: Applied Physics* 39: 21–24.
24. Sekita, M., M. Nakamura, A. Watanabe, S. Takekawa, and K. Kitamura. 2006. Induced emission cross sections of near-stoichiometric $LiNbO_3$: Mg, Nd. *Journal of Applied Physics* 100: 103501.
25. Jaquel, D., J. Capmany, J. Garc'ia Sol'el, A. Brenier, and G. Boulon. 2000. Continuous-wave laser properties of $^4F_{3/2} \to {}^4I_{13/2}$ channel in the $Nd^{3+}LiNbO_3$: ZnO non-linear crystal. *Applied Physics B: Lasers and Optics* 70: 11–14.
26. Kitamura, K., J. K. Yamamoto, N. Iyi, S. Kimura, and T. Hayashi. 1992. Stoichiometric $LiNbO_3$ single crystal growth by double crucible Czochralski method using automatic powder supply system. *Journal of Crystal Growth* 116: 327–332.
27. Bordui, P. F., R. G. Notwood, D. H. Jundt, and M. M. Fejer. 1992. Preparation and characterization of off-congruent lithium niobate crystals. *Journal of Applied Physics* 71: 875–879.
28. Solanki, S., T. Chong, and X. Xu. 2003. Flux growth and morphology study of stoichiometric lithium niobate crystals. *Journal of Crystal Growth* 250: 134–138.
29. Tsai, C. B., W. T. Hsu, M. D. Shih, Y. Y. Lin, Y. C. Huang, C. K. Hsieh, W. C. Hsu, R. T. Hsu, and C. W. Lan. 2006. Growth and characterizations of ZnO-doped near-stoichiometric $LiNbO_3$ crystals by zone-leveling Czochralski method. *Journal of Crystal Growth* 289: 145–150.
30. Kumaragurubaran, S., S. Takekawa, M. Nakamura, S. Ganesamoorthy, K. Terabe, and K. Kitamura. 2005. Domain inversion and optical damage in Zn doped near stoichiometric lithium niobate crystal. Paper presented at the Conference on Lasers and Electro-Optics (CLEO), CMW2, p. 393.
31. Babu Reddy, J. N., K. Ganesh Kamath, S. Vanishri, H. L. Bhat, and E. J. Suja. 2008. Influence of Nd: Zn codoping in near-stoichiometric lithium niobate. *The Journal of Chemical Phsyics* 128: 244709, 465–470.
32. Kolm, C., and P. Moody. 1958. Syringe-type single-crystal furnace for materials containing a volatile constituent. *Review of Scientific Instruments* 29: 1144–1145.
33. Baldwin, E. M. N., J. C. Brice, and E. J. Millett. 1965. A syringe crystal puller for materials having a volatile component. *Journal of Scientific Instruments* 42: 883–884.

34. Gremmelmaier, R. 1956. Herstellung von InAs- und GaAs-Einkristallen. *Zeitschrift für Naturforschung* 11a: 511–513.
35. Beck, A., and E. Mooser. 1961. Apparaturzum Ziehen von Einkristallen von ... mitleichtfluchtigen Komponenten. *Helvetica PhysicaActa* 34: 370–373.
36. Metz, E. P. A., R. C. Miller, and R. Mazelsky. 1962. A technique for pulling single crystals of volatile materials. *Journal of Applied Physics* 33: 2016–2017.
37. Bass, S. J., and P. E. Oliver. 1968. Pulling of gallium phosphide crystals by liquid encapsulation. *Journal of Crystal Growth* 3(4): 286–290.
38. Stepanov, V. A. 1959. A new technique of production of sheets, of tubes of rods with different cross sections from a melt. *Soviet Physics* JETP 29: 382–393.
39. Stepanov, V. A. 1959. A new technique of products fabrication from a melt. *Bulletin of Mechanical Engineering* 11: 47–50.
40. Tatarchenko, V. A. 1994. Shaped crystal growth. In *Handbook of crystal growth*, Vol. 2, edited by D. T. J. Hurle. Amsterdam: North Holland, pp. 1015–1110.
41. Antonov, P. I., and V. N. Kurlov. 2002. New Advances and Developments in the Stepanov Method for the Growth of Shaped Crystals. *Crystallography Reports* 47(1): S43–S52.
42. Theodore, F., T. Duffar, and J. N. Santailler. 1999. Crack generation and avoidance during the growth of sapphire domes from an element of shape. *Journal of Crystal Growth* 204: 317–324.
43. La Belle, H. E., and A. J. Mlavsky. 1967. Growth of sapphire filaments from the melt. *Nature* 216: 574–575.
44. La Belle, H. E., and A. J. Mlavsky. 1971. Growth of controlled profile crystals from the melt: Part I—Sapphire filaments. *Materials Research Bulletin* 6: 571–579.
45. La Belle, H. E. 1971. Growth of controlled profile crystals from the melt: Part II—Edge-defined, film-fed growth (EFG). *Materials Research Bulletin* 6: 581–589.
46. Fukuda, T., P. Rudolph, and S. Uda, eds. 2004. *Fiber crystal growth from the melt*. Berlin, Germany: Springer.
47. Fukuda, T., and V. I. Chani. 2007. Part I: Introduction to micro- pulling down method. In Shaped crystals: *Growth by micro-pulling-down technique*, edited by T. Fukuda and V. I. Chani. Berlin, Germany: Springer-Verlag, pp. 3–92.
48. Kapitza, P. 1928. The study of the specific resistance of bismuth crystals and its change in strong magnetic fields and some allied problems. *Proceedings of the Royal Society of London A* 119: 358–443.
49. Pfann, W. G. 1952. Principles of zone melting. *Transactions of the American Institute of Mining and Metallurgical Engineers* 194: 747–753.
50. Pfann, W. G. 1958. *Zone melting*. New York, NY: John Wiley.
51. Keck, P. H., and M. J. E. Golay. 1953. Crystallization of silicon from a floating liquid zone. *Physical Review* 89: 1297–1297.
52. Burton, J. A., R. C. Prim, and W. P. Slichter. 1953. The distribution of solute in crystals grown from the melt. *Journal of Chemical Physics* 21: 1987–1991.
53. Andrade, C. E. N., and R. Roscoe. 1937. Glide in metal single crystals. *Proceedings of the Physical Society of London* 49: 152–177.
54. Schildknecht, H. 1966. *Zone melting*. New York, NY: Academic Press.
55. Heywang, W. 1956. Zur Stabilität senkrechter Schmelzzonen *Zeitschrift fur Naturforschung* 11a: 238–343.
56. Heywang, W., and G. Ziegler. 1954. Zur Stabilität senkrechter Schmelzzonen. *Zeitschrift fur Naturforschung* 9a: 561–562.

57. Harikrishnan, S. 2009. Phase transition and magnetic order in multiferroic and ferromagnetic rare earth manganites. PhD dissertation, Indian Institute of Science, Bangalore, India.

58. Ivanov, V. Y., A. A. Mukhin, A. S. Prokhorov, A. M. Balbashov, and L. D. Iskhakova. 2006. Magnetic properties and phase transitions in hexagonal $DyMnO_3$ single crystals. *Physics of the Solid State* 48: 1726–1729.

59. Alonso, J. A., M. J. Martinez-Lope, M. T. Casais, and M. T. Fernandez-Diaz. 2000. Evolution of the Jahn–Teller distortion of MnO_6 octahedra in $RMnO_3$ perovskites (R = Pr, Nd, Dy, Tb, Ho, Er, Y): A neutron diffraction study. *Inorganic Chemistry* 39: 917–923.

60. Verneuil, A. 1902. Production artificielle du rubis par fusion. *Comptus Rendus* 135: 791–794.

61. Falckenberg, R. 1978. The Verneuil process. In *Crystal growth: Theory and techniques*, Vol. 2, edited by C. H. L. Goodman. New York, NY: Plenum Press, pp. 109–184.

62. Verneuil, A. 1904. Mémoire sur la reproduction artificielle du rubis par fusion. *Annals of Chemistry Ser 8* T3: 20–48.

63. Merker, L. 1947. Remembrances of flame fusion. In *50 Years progress in crystal growth: A reprint collection*, edited by R. S. Feigelson. Amsterdam, The Netherlands: Elsevier, pp. 23–28.

64. Merker, L. 1962. Synthesis of calcium titanate single crystals by flame fusion technique. *Journal of the American Ceramics Society* 45: 366–369.

65. Mitchell, R. S. 1965. After-heater furnace for verneuil crystal growing technique. *Review of Scientific Instruments* 36: 1667–1668.

66. Alexander, A. E. 1949. The synthesis of rutile and emerald. *Journal of Chemical Education* 26: 254–254.

67. Popov, S. K. 1946. Novyi proizvodstvennyi metod vyrashchivaniia kristallov korunda. *Bulletin of the Academy of Sciences USSR: Division of Chemical Science* 10: 504–508.

68. Scheel, H. J., J. G. Bednorz, and P. Dill. 1976. Crystal growth of strontium titanate $SrTiO_3$. *Ferroelecrrics* 13: 507–509.

69. Falckenberg, R. 1975. Growth of Mg-Al spinel crystals of large diameter using a modified flame fusion technique. *Journal of Crystal Growth* 29: 131–224.

70. Lefever, R. A., and G. W. Clark. 1962. Multiple-tube flame fusion burner for the growth of oxide single crystals. *Review of Scientific Instruments* 33: 769–770.

71. Halden, F. A., and R. Sedlacek. 1963. Verneuil crystal growth in the arc-image furnace. *Review of Scientific Instruments* 34: 622–626.

72. Rabenau, A. 1964. Herstellung von MgO—Einkristallen. *Chemical Engineering & Technology* 36: 542–545.

73. Schupp, L. J. 1968. Magnesia crystals grown in carbon arc furnace. *Electrochemical Technology* 6: 219–221.

74. Fan, J. C., and T. B. Read. 1972. Growth of crystals of V_2O_3 and $(V_{1-x}Cr_x)_2O_3$ by the tri-arc Czochralski method. *Materials Research Bulletin* 7: 1403–1409.

75. Osiko, V. V., M. A. Borik, and E. E. Lomonova. 2010. Synthesis of refractory materials by skull melting techniques. In *Springer handbook of crystal growth*, edited by G. Dhanaraj et al. New York, NY: Springer, pp. 443–477.

76. Kuzminov, Y. S., E. E. Lomonova, and V. V. Osiko. 2008. *Cubic Zirconia and skull melting*. Cambridge, UK: Cambridge International Science.

9

Growth from Liquid Solutions

Introduction

Growth from liquid solutions falls into the category of polycomponent growth techniques. Here, there will be at least two components—namely, the solute and the solvent. Additives may be added to enhance the solubility. Growth ambience may also be changed for the same purpose. When compared with growth from melt, crystal growth from solution results in remarkable improvements in the quality of the grown crystals. This is principally due to the much lower crystallization temperature. Lower temperature would mean lower number density of structural defects and less contamination from the container material. Also, the low growth rates involved in the low-temperature process enable better control over various growth parameters, which are crucial for successful crystal growth.

The methods of growing crystals from liquid solution may be further divided into the following five types:

1. Growth from aqueous solution
2. Growth from silica gel
3. Hydrothermal growth
4. Growth from flux
5. High-pressure growth

In the following sections, we discuss these techniques one by one, but we will limit our discussion mainly to the basics.

Growth from Aqueous Solution

As indicated earlier, the growth from aqueous solution is probably the oldest method of crystal growth. A large number of industrial processes forming crystalline products ranging from table salt to complex pharmaceuticals employ this technique. However, in all these processes, the main objective is to obtain a high yield of uniform-sized crystalline product where the size required is quite small and the crystalline perfection is of no importance. Here, the growth is either by spontaneous nucleation or by seeding where the seeds are crystalline powders. Such processes are of no use in growing large and perfect crystals.

One of the greatest advantages of crystal growth from solution is the control and stability it provides on the temperature of growth. This makes it possible to grow crystals that exist in several crystal forms, depending on the temperature. A second advantage is the control of viscosity that one can exercise (by varying the solution temperature), thereby facilitating growth of materials that tend to form glasses on cooling from their melts. Further, crystals grown from solutions usually have well-defined habit faces. This enables the identification of the crystallographic axes with respect to the external shape. Also, the concentration of many compounds in liquid solutions can be quite high, resulting in a high yield on crystallization.

The classic book on crystal growth by Buckley [1] is very instructive for those who would like to pursue solution growth technique seriously. The book by Holden and Singer [2], though at an introductory level, would be inspiring for beginners.

Solution and Solubility

By definition, a solution is a homogeneous mixture of two or more substances; it may be gaseous, liquid, or solid, and its components are usually called solute and solvent. However, we limit our discussion only to those solutions in which solvent is a liquid and solute a solid. Although there is no particular reason as to why one particular component of a solution should be termed the solvent, it is conventional to call the component present in excess as solvent. Many exceptions, however, exist to this convention.

The composition of a solution can be expressed in a number of ways. The amount of solute present in a solution is most conveniently stated as parts by weight of solute per part by weight of solvent. Solubility is thus defined as the number of grams of the solute dissolved in 100 g of the solvent. To avoid confusion in the case of hydrated salts dissolved in water, the solute concentration should always refer to the anhydrous salt.

When table sugar is added to water at room temperature, say 30°C, about 220 g of it dissolves in 100 g of water to form a homogeneous solution. Further addition of sugar leads to the sugar remaining as a solid precipitate at the bottom of the flask and in contact with the liquid. Such a solution is termed as saturated at that temperature as it will take in no more of the solid. We may view this situation as an equilibrium situation.

If we were to change the temperature of the system, the amount of the solute in solution will generally vary. In the case of sugar, its solubility in water is known to increase with temperature because the dissolution is endothermic here. Other examples belonging to this category are benzoic acid, KNO_3, $NaNO_2$, KH_2PO_4, and so on. On the other hand, dissolution of Li_2SO_4, Li_2CO_3, $MnSO_4$, and Na_2SO_4 is exothermic; hence, for them the solubility decreases with increase in temperature.

The solubility characteristics of solute–solvent systems have considerable influence on the choice of the method of crystallization. It would be useless, for instance, to cool a hot saturated solution of NaCl in the hope of growing crystals of any quantity. Cooling from, say, 90°C to 20°C would produce only about 3 g of NaCl for every 100 g of water present. The yield could be improved, however, by removing some water by evaporation, and this is what is done in practice. On the other hand, a direct cooling operation would be adequate for a salt such as $CuSO_4$ because its solubility varies steeply with temperature. Cooling from 90°C to 20°C would produce 48 g of $CuSO_4$ for every 100 g of water.

As illustrated in Chapter 6, not all solubility curves are smooth. A discontinuity in the solubility curve denotes a phase change. The solubility curves of two different phases meet at the transition point, and a system may show a number of such points depending on the number of phases present. All these issues have to be considered while growing crystals from solution.

As the solubility curves indicate, at a given temperature a solvent can contain a fixed amount of solute to make the solution saturated. However, it is not very difficult to prepare a solution containing more solid than that represented by saturation conditions, and such a solution is said to be supersaturated. Uncontaminated solution in clean containers cooled slowly without disturbances and in dust-free atmosphere can readily be made to show appreciable degrees of supersaturation. As all of us know, the state of supersaturation is an essential condition for crystallization. Supersaturation can also be achieved by removing some of the solvent from the solution by evaporation. However, in practice, a combination of cooling and evaporation is employed.

Choice of Solvent

The selection of the best solvent for the growth of a given material is not always easy. Many factors have to be considered, and some compromise must invariably be made. Broadly speaking, a good solvent should have the following characteristics: (1) moderate solubility with a reasonable positive slope in the solubility curve, (2) low volatility, (3) nontoxicity, (4) noncorrosiveness, (5) noninflammability, and (6) low cost. In addition, the solvent should be stable under all possible operating conditions. For example, it should not undergo chemical changes such as oxidation or decomposition. It should also not react chemically with the solute. Probably, no solvent exists with all these attributes. In practice, one may have to accept several undesirable characteristics of a solvent if that solvent has one (or two) important solvent property that is crucial for growth. A solvent is preferred in which the solute is soluble to the extent of 10%–60%. In the case of volatile solvent, precautions must be taken to prevent evaporation because fast evaporation promotes spurious nucleation, particularly at the free surface due to temperature and concentration changes. Also, close chemical similarity between the solute and the solvent must be avoided because their mutual solubility might render the crystallization difficult.

Probably no other solvent is as generally useful for growing crystals as water. Its high solvent action, which is related to its high dielectric constant, high stability, low viscosity, low toxicity, and its availability in pure state are some of the properties that account for this. Also, water offers a very good control of supersaturation by variation of temperature or controlled removal of solvent. The book by Franks on water [3], which discusses its remarkable properties, its influence on dissolved substances, and its controlling role in the life sciences and ecology, is recommended for further reading.

There are, however, many organic liquids that are capable of acting as solvents. With the discovery of many organic crystals that are potentially applicable in nonlinear optics, they have become all the more important. Prominent among them are acetic acid and its esters, lower alcohols and ketones, ethers, chlorinated hydrocarbons such as $CHCl_3$ and CCl_4, benzenes, toluene, xylene, and so on. Often, a mixture of two or more solvents has been found more suitable for the crystallization process than the individuals. Common binary solvent mixtures that have proved useful are water–alcohol, alcohol–ketones, alcohol–ether, alcohol–CCl_4, alcohol–$CHCl_3$, alcohol–benzene, toluene–xylene, benzene–hexane, and hexane–trichloroethylene.

If the solubility of a material in a given solvent is too small to be useful, one can try to improve it by adding a small quantity of complex-forming agent. Mineral acids, alkalis, and sulfides are some of the agents found useful. Addition of acids and alkalis also alters the pH value of the solution, which is found to have a profound effect on the growth rate.

There is a frequently quoted rule "like dissolves like" that can serve as a useful guide in selecting a solvent, although there are many exceptions to it. For example, solvents may be either polar or nonpolar. The polar liquids (e.g., water) usually have high dielectric constants, and the nonpolar (e.g., benzene) have low dielectric constants. A nonpolar solute is generally more soluble in a nonpolar solvent (e.g., anthracene in benzene) than a polar solvent (water). However, one should not blindly apply this rule.

Additives

It is an established fact that a small quantity of some additives or impurities when added to the growing system will affect the growth remarkably. There are a wide variety of additives that influence the crystal growth and morphology. Additives may increase the growth rates by affecting the activation energy of nucleation of growth layers. In fact, the face-adsorbed foreign molecules affect the growth either by favoring the elementary process of attachment of atoms/molecules by decreasing the amount of energy required for the process or by retarding their surface diffusion on the adsorbed layer. It is the relative weighting of these processes that decides the growth rate of the layer. During growth, if an impurity ion can attach itself to a high-energy site, it would interrupt the growth sequences, thereby causing the reduction in the growth rate. It should be remembered that the relative changes in the growth rates of various faces lead to habit modification. Finally, we may note that the impurities need not necessarily get into the growing crystal.

The effect of impurities can often be exploited to the practical advantage of altering crystal morphology. Mullin et al. [4] have reported that the presence of foreign ions such as Cr^{3+}, Fe^{3+}, and Al^{3+} in low concentrations (~50 ppm) causes a tapering in ammonium dihydrogen phosphate (ADP) and potassium dihydrogen phosphate (KDP) crystals. Perhaps, the classic example of an impurity that promoted the growth of good quality crystals is Pb^{2+}, without which NaCl crystal would grow as cloudy hydrate [5]. Any number of such examples can be found in the literature. With ionic crystals, it is frequently desirable to put ionic impurities into the crystal in order to control the electronic properties. In doing so, the relative ionic radii of the host and the guest ions and the charge balance in the crystal have to be considered.

Nucleation

Perhaps, the best nucleus for the growth of a single crystal is a small seed crystal of the same material itself. However, there is much evidence that other materials of similar crystal structure and perhaps even the replicas of crystal faces will initiate crystal growth. It is usually advantageous to select a seed crystal of such an orientation that would result in the growth along only one or two directions. To accomplish this, a seed crystal of desired cross-sectional area and orientation has to be produced. An increase in cross-sectional area can sometimes be obtained by forcing the crystals to grow at near maximum rate by the use of additives. It is a fact that the general tendency of a crystal is to have a habit bounded by faces of low Miller indices. Since the high-index planes grow faster, seed plates cut parallel to such planes grow at faster rates. A significant increase in the growth rate can be achieved by using seed plates whose surfaces are not parallel to natural faces. It has been observed that growth on such surfaces (in the case of KDP or ADP) would be very fast until the natural faces are restored [6]. This process is referred to as capping.

Achievement of Supersaturation

The most critical factor in the crystal growth process is probably the achievement of supersaturation. The following three methods are generally used:

1. The temperature of the saturated solution is raised or lowered depending on the sign of the temperature coefficient of solubility.
2. A temperature gradient is established in the solution, resulting in continuous dissolution of excess nutrient in one zone and continuous extraction by crystal growth in a second zone.
3. A solution contained in one vessel at a constant temperature is continuously saturated with respect to that temperature and then pumped into a second vessel that contains a seed crystal at another temperature (usually lower).

In all the aforementioned methods, the supersaturation is achieved by exploiting the temperature coefficient of solubility. However, as mentioned earlier, supersaturation can also be achieved by evaporation of the solvent from a solution kept at constant temperature (i.e., by solvent evaporation method).

Apparatus

With the earlier prescriptions, a host of low-temperature solution growth apparatuses have been designed and fabricated over the years. Here we discuss only a few classic cases.

Mason-Jar Method

A very simple technique for slow cooling is the Mason-jar method, elegantly described by Holden and Singer [2]. Because it does not require any sophisticated apparatus, it is quite cheap; hence, it facilitates crystal growth in batches. What is required here is something like a kitchen jar with a screw cap, often called a Mason jar. These jars were invented and patented in 1858 by Philadelphia tin smith John Landis Mason (hence the name). Initially, a saturated solution is prepared at a temperature that is a few degrees above the room temperature. This solution is heated slightly above its saturation temperature to make it undersaturated and poured into this jar. Then, a seed tied to one end of a piece of silk thread, the other end being fixed to the screw cap, is introduced by screwing the cover onto the jar,

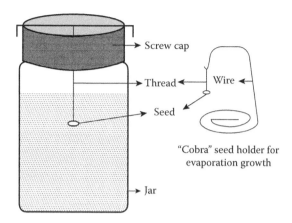

FIGURE 9.1
Crystal growth using a Mason jar.

as shown in Figure 9.1. The jar is left undisturbed and allowed to cool to room temperature by natural cooling. If the solubility of the chosen material is high and the system is allowed to cool slowly, reasonably good crystals can be obtained by this method. The required seeds can be obtained by spontaneous nucleation by simply allowing a saturated solution to evaporate slowly. In Mason-jar technique, the cooling rate, of course, is uncontrolled and no stirring of the solution is possible. Nevertheless, the simplicity of the method still makes it attractive. Consequently, the technique is widely used in one form or the other in crystal growth laboratories, particularly for preliminary studies.

The Mason-jar method may easily be adapted to grow crystals by solvent evaporation. The procedure is as described earlier, except that the seed is suspended from a seed holder that is in the form of a "wire cobra" for stability and is immersed in the solution (see Figure 9.1). Since this seed holder does not intersect the solution surface, it will not provide sites for nucleation at the interface, which gets highly supersaturated during evaporation. To facilitate evaporation, the jar is covered with a cloth fastened at the lip with a rubber band. Otherwise, a number of small holes can be drilled in the screw cap. The initial growth will be by slow cooling, but once the solution reaches room temperature, subsequent growth would occur by solvent evaporation. If the evaporation has to take place at a higher rate, the jar may be heated with the help of a hot plate or water bath. A room maintained at constant temperature is generally desirable for such growth.

If excess solute is kept on the floor of the jar and heated at the bottom by a hot plate to maintain it at a slightly higher temperature, a temperature gradient will be set up between the bottom and top. This facilitates dissolution in the lower zone and growth in the upper zone, where the seed is placed.

Holden's Rotary Crystallizer

For growth by slow cooling, the most useful apparatus is Holden's rotary crystallizer [7]. Figure 9.2 shows a schematic drawing of a version of this apparatus. Initially, the growth vessel, often made of glass, is filled with saturated or near-saturated solution with respect to the initial growth temperature. To avoid spurious nucleation, it is advantageous to transfer the solution at a temperature slightly above its saturation temperature. Alternately, the solution can be prepared in the vessel itself, provided that the solute is free from suspended particles. Suitably oriented seeds are mounted on the seed holder. As mentioned earlier, seeds are usually prepared by solvent evaporation technique. Seed holders for the Holden's crystallizer are constructed from materials like glass, poly-tetrafluoroethylene (PTFE), stainless steel, tungsten-rod stock, and so on. It is often possible to check for saturation of an aqueous solution by inserting a small test seed near the surface and observing the movement of Schlieren lines caused by refractive index variations. These variations are associated with the concentration gradients that exist around the seed crystal when it is not in equilibrium with its solution. If these lines appear to be descending from the seed, the liquid around it is denser than the surrounding

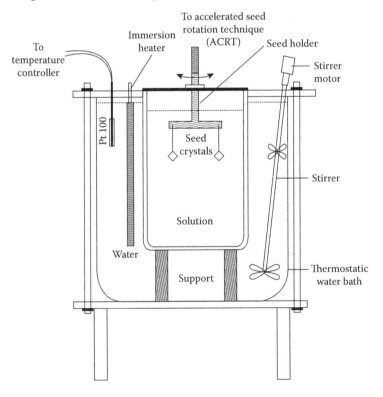

FIGURE 9.2
Holden's rotary crystallizer.

solution and the crystal is dissolving, which means the solution is undersaturated. On the other hand, if the lines are rising, more dense fluid is approaching the seed crystal from the solution; hence, the solution is supersaturated. Depending on the results of this test, the temperature of the solution may be adjusted to bring the system as close to equilibrium as possible. Once the test seed attains equilibrium, it is withdrawn and the actual seed crystal is inserted to commence the growth by cooling down the solution.

Temperature of the solution is maintained at the desired level by a variety of heater controller arrangements. Usually, solution flask is immersed in a temperature-controlled water bath. The water bath is heated by resistance heaters immersed in the bath. More rapid response is obtained when an infrared lamp is used to heat the water bath surrounding the solution. Temperature-sensing elements may be as simple as mercury thermo regulators, although more sophisticated ones are used. Volume of the water bath depends on the size of the solution flask, whose size in turn depends on the size of the crystal to be grown. It is usually customary to slowly rotate the seed holder to decrease the diffusion effects. To prevent the formation of vortices, it is generally common to reverse the rotation direction every few seconds. The top cover of the flask is generally sealed as tightly as possible to prevent solvent evaporation, and the seed holder is passed through some seal (Teflon or mercury seal are common) that allows rotation of the seed holder.

Supersaturation is provided by slow cooling, and the cooling rate is determined by the system. Usually, cooling rates of 0.1°C–1.5°C/day or sometimes even slower are found in the literature. Motor-driven, cam-type controllers have been used to set the cooling rate, but one can also resort to setting the control point of the thermo regulator by a predetermined amount once each day. Clearly, crystal growth by slow cooling is not isothermal. Hence, such properties of the crystal that depend on the growth temperature will not be uniform throughout.

Modifications

There have been many variants of Holden's rotary crystallizer, with these modifications arising essentially because of the requirements of the individual crystal growers. Figure 9.3 shows a schematic of the apparatus built in the author's laboratory along with its photograph [8]. This apparatus is similar in many ways to the one developed by Hooper et al. [9], except for some minor variations due to local constraints. Referring to the schematic diagram, the solution is contained in a sealed central flask (F) with a greaseless vacuum seal and an attached seed holder rod (GR) with Teflon seal. The solution of the central flask is heated by the external 20 L bath (B), which contains water that is being continuously stirred (S) for temperature uniformity. The large volume of the bath prevents rapid temperature fluctuations and hence ensures good temperature controlling. The thermostatic bath is heated with an infrared lamp (L) at the base of the unit, and the control

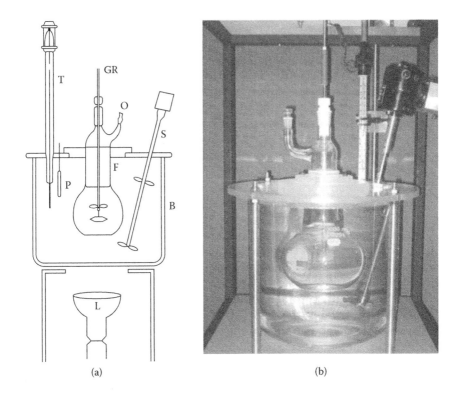

(a) (b)

FIGURE 9.3
Modified Holden's rotary crystallizer: (a) schematic and (b) photograph.

is achieved by the adjustable long-range electrical contact thermometer (T). The internal solution is stirred by the paddles attached to the seed holder, which rotates at a speed of 1 rev/10 s, alternatively in clockwise and anti-clockwise directions to ensure continuous and efficient agitation of the solution. The supersaturation of the solution is achieved by driving the contact of the electrical contact thermometer to lower temperature using a syn-chronous motor attached to the magnetic setting device. The rate at which the temperature is lowered can be varied by a cam-operated, time-sharing device, which switches the contact thermometer motor drive "on" for vary-ing periods of time, depending on the disposition of the cam relative to the microswitch. Using a 1 rev/h motor to drive the contact electrical thermom-eter and a 1 rev/min motor to drive the cams and changing the disposition of cams appropriately, linear lowering rates in the range of 0.1°C–1.0°C/day could be obtained.

Slow evaporation of the solvent at a constant temperature can most easily be achieved by providing an opening (O) on the top of the solution flask, as shown in the diagram, and the evaporation rate is controlled by the relative size of the opening available for the solvent to escape. This method of solvent

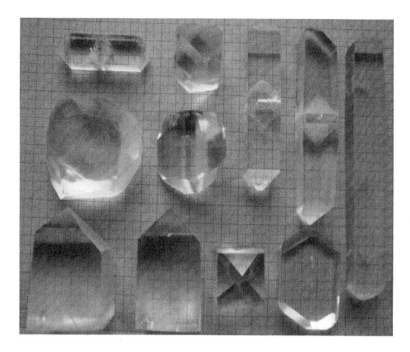

FIGURE 9.4
Representative crystals grown using the crystallizer shown in Figure 9.3.

evaporation may not be precise, but it works satisfactorily in many cases. The rate of removal of the solvent vapor may be more precisely controlled by a controlled flow of air through the flask by providing both an inlet and an outlet.

The system just described is versatile, efficient, easy to operate, and above all quite cheap. Since both the central solution flask and thermostatic water bath are made from glass, the growing crystal can be seen easily. This facilitates midcourse correction and intervention in case of necessity. The system over the years has seen many improvements, and currently the electromechanical temperature-controlling system has been replaced by the commercial Eurotherm programmable temperature controller with Pt 100 sensor (P). Figure 9.4 shows representative crystals grown using this apparatus with appropriate solution flasks.

Temperature Differential Method

Growth by the temperature differential method was first reported by Kruger and Finke [10]. A modified version of their apparatus is shown schematically in Figure 9.5. This is a two-vessel system. In one of them, called the dissolution vessel, maintained at a temperature of say T_1, excess solute is placed at the bottom. The solution in it is continuously stirred to keep it always

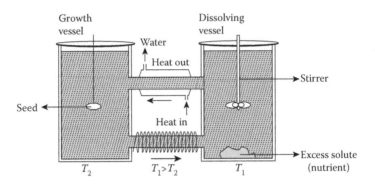

FIGURE 9.5
Temperature differential method.

saturated with respect to temperature T_1. This solution is sent to the growth vessel, which is maintained at a temperature $T_2 < T_1$. The growth vessel contains a seed holder from which a seed crystal is suspended. The saturated solution on reaching the growth vessel becomes supersaturated because of the temperature difference and causes the seed to grow. The depleted solution returns to the dissolution vessel, where it becomes undersaturated and hence dissolves the excess solute present in it. Hence, through the solution circulation, the excess nutrient present in the dissolution vessel is transported to the seed crystal, where it gets deposited. In a typical growth run, initially both the growth and dissolving vessels are brought to the same temperature ($T_1 = T_2$) and held at that temperature until the solution in the growth vessel gets saturated. Then, the temperature in the growth vessel is slightly increased ($T_2 > T_1$) so that solution becomes slightly undersaturated. Now when a seed is inserted, the surface damages, if present, are removed by dissolution. Usually the seed is rotated or oscillated as in a rotary crystallizer. Subsequently, the temperature T_2 is slowly brought down until $\Delta T = T_1 - T_2$ is of the desired value. The circulation of the solution between the dissolution and growth vessels is affected by convection via the two connecting tubes (see Figure 9.5). These tubes essentially serve as heat exchangers. By winding the heating tape around the lower tube and providing a cold water jacket around the upper tube, one may further promote the exchange process. A peristaltic pump that squeezes plastic tubing connecting the two vessels rhythmically to cause liquid motion will promote solution circulation. A filter in the upper tube, if inserted, helps to prevent small undissolved crystallites, if any, from entering the growth vessel.

Walker and Kohman [11] have described a three-vessel growth system for producing large-sized ADP and EDT crystals by the temperature differential method. In this system, tanks 1 and 2 have the same role as those in the conventional two-vessel system. The third tank, called the superheater, heats the solution to make it undersaturated. This hot solution is pumped into the growth vessel from the bottom through a heated pipeline. The details can be found in Ref. 11.

It must be mentioned here that in the slow-cooling method, the crystal actually grows at a varying temperature, which means different parts of the crystal grow at different temperatures. In the solvent evaporation technique, since the solution is usually kept at a constant temperature, the crystal grows at a fixed temperature. However, the evaporation rate depends on many factors and may vary during the growth, thereby changing the degree of supersaturation. This would lead to variation in growth rate during the growth period. Both these problems are overcome in the temperature differential method in which the crystal grows not only at constant temperature but also at constant supersaturation (constant ΔT). Consequently, this technique has been considered as the most suitable one for the growth of large-sized crystals.

Growth by Fast Crystallization

One of the limitations of solution growth described earlier is the rate of growth. Growth rates reported in the literature are normally in the range of 1–2 mm/day, which are quite low when compared to the rates available with melt-growth techniques, which can be as high as 20 mm/h. This implies that the supersaturation levels used in these growth experiments are low. As a matter of fact, a high supersaturation required for large growth rates can neither be achieved nor be maintained in conventional Holden-type rotary crystallizers. This is essentially because of the frequent occurrence of spontaneous nucleation during the extended growth period. Consequently, growth of large crystals from aqueous solution had remained as a challenge to the crystal growers.

However, in recent times, there has been greater demand for large-sized KDP and related crystals, the sizes of which are 50–60 cm in lateral dimensions. These crystals are essentially used in the area of high-power laser technology, such as National Ignition Facility, Lawrence Livermore National Laboratory, California, for nuclear fusion related research. Growth of such large crystals by the conventional method would take approximately 2 years, resulting in higher costs due to the long-term labor and machine running. To meet this challenge, a new line of activity started in the later part of the last century, and this led to the development of fast-growth techniques [12–14]. The fast crystallization technique is also referred to as a platform technique by some scientists. Although much of the research and development in this area has been directed toward the growth of large-sized KDP crystal and its deuterated analog [15], other crystals such as lithium iodate, ADP, cesium dihydrogen arsenate, l-arginine phosphate, and others have also been researched.

The two most important requirements for the rapid crystal growth are higher level of supersaturation and suppression of spontaneous crystallization in the mother solution. However, at such high supersaturation levels, solution stability depends on many parameters such as overheating of the solution, purity of raw materials, crystallizer design, pH, hydrodynamic conditions, rate of temperature reduction, and so on. These have been discussed in detail in the literature [12–18]. We describe here a laboratory-scale setup that was used successfully to grow KDP crystals by the rapid crystallization technique.

The experimental arrangement for growing crystals by the rapid crystallization process is schematically shown in Figure 9.6. Referring to the figure, the solution is contained in a 5 l cylindrical flask with a lid to which Teflon platform is anchored through a ball bearing. The internal solution is stirred by this Teflon platform at a speed ranging from 150 to 300 rpm. For the favored global mixing of the solution in radial and axial directions during growth, the design of the platform is modified, particularly with regard to the number of vertical posts (initially three and then four). Stirring of the solution achieved by the accelerated seed rotation technique also helps to increase the growth rate in lateral directions as well as change in morphology of the crystal because of forced solution convection. The solution of the cylindrical flask is heated by the external 20 L water bath,

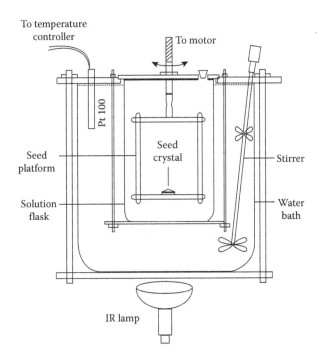

FIGURE 9.6
Crystallizer shown in Figure 9.3 modified to accommodate a platform as a seed holder and a larger cylindrical flask.

similar to the one used in the modified Holden's rotary crystallizer shown in Figure 9.3 with its heating and temperature-controlling arrangements.

On an experimental basis the growth of KH_2PO_4 (KDP) single crystal was carried out using this setup. The solution was prepared using triple-distilled water obtained by distilling deionized water. Analytical Reagent (AR) grade KH_2PO_4 (KDP) material was purified by repeated crystallization before preparing the mother solution. This process was followed principally to remove the metal ion impurities such as Al, Fe, and Cr, which can easily get incorporated into the prismatic sectors and deteriorate the optical quality of the crystals. For removing suspended impurities, the solution was subjected to the centrifuge technique at 15,000 rpm. The resulting solution was filtered with a micropore filter and transferred to the crystallizer for growth. In a typical growth run, the range of growth temperature was 45°C–36°C, the cooling rate was 3°C/day, and the rotation rate of the platform was 150 rpm (max). The maximum growth rate achieved along the *a/b* axis was 7–7.5 mm/day. With these conditions KDP crystals up to $40 \times 43 \times 66$ mm^3 size could be grown in about 3 days. x-Ray and optical characterization carried out on these crystals revealed no significant difference between the crystals grown by the rapid crystallization process and the traditional method. Figure 9.7 [18,19] shows two crystals grown by this platform technique.

To grow large crystals of KDP required by the National Ignition Facility, the laboratory used a 6-foot-high tank filled with nearly a thousand liters of supersaturated KDP solution. The crystal grew by the fast crystallization method on a rotating platform, and in just 2 months, it grew to an approximate size of $45 \times 45 \times 45$ cm^3, weighing nearly 365 kg. With conventional methods of growing crystals, this process would take nearly 2 years.

(a)   (b)

FIGURE 9.7

(a) KDP crystal growing on a platform and (b) Deuterated Potassium Dihydrogen Phosphate (DKDP) crystal grown by the platform technique with a point seed. (Panel a is from Dixit, V. K., et al., *Bulletin of Materials Science*, 24, 455–459, 2001. With permission. Panel b is from Karnal, A. K., Growth and characterization of technologically important nonlinear optical crystals cesium lithium borate and potassium di deuterium phosphate, PhD dissertation, Indian Institute of Science, Bangalore, India, 2006.)

TABLE 9.1

Growth of Large-Sized KDP Crystals

Method of Growth	Conventional Temperature-Lowering Method	Fast Crystallization Method
Volume of the solution	4000 L	1000 L
Material transport	Crystal oscillation	Solution circulation and crystal oscillation
Range of growth temperature	60°C–30°C	60°C–30°C
Cooling rate	0.1°C/day (average)	1.0°C/day (average)
Supersaturation range	—	16%–6%
Growth rate	1–2 mm/day along c	7–8 mm/day along a and b 15 mm/day along c
Duration of growth	~300 days	~30 days
Final dimensions of the grown crystal	40 × 40 × 60 cm³	45 × 45 × 45 cm³
Reference	[20]	[14]

Table 9.1 provides a comparison between the conventional and fast crystallization methods in terms of various growth parameters used for growing large KDP crystals. These parameters were extracted from two papers published in *Journal of Crystal Growth* [14,20].

Growth from Silica Gel

Growing crystals in gels has been known since the end of the nineteenth century, but the method went into a state of hibernation until the interest in it was revived afresh by Henisch and his coworkers through their extensive work in the 1960s [21]. The method was initially used to grow only those crystals that had very low solubility in water. Since then, various modifications have been introduced to the basic gel method, and different types of crystals, including the ones soluble in water, have been grown with a high degree of perfection. Although the size of the crystals that could be grown by the gel method is limited, its ability to grow practically insoluble materials (in water) at room temperatures with exceedingly simple and inexpensive equipment makes the technique attractive. Also, since gel medium is almost transparent, the growing crystals can be viewed during the entire growth period. Further, since the growth occurs at room temperature, one can expect them to contain a lower concentration of equilibrium defects than those grown by high-temperature techniques.

The developments up to 1986 have been well documented in the revised book written by Henisch [22]. Subsequently, Lefaucheux and Robert [23] reviewed the work in this field in Chapter 20 of the *Hand Book of Crystal Growth*. The recent interest in gel growth has, however, been on account of its suitability to grow crystals of biological macromolecules and studies related to biomineralization, some aspects of which have been reviewed by Kalkura and Natarajan [24].

In this method, one reagent is incorporated in the gelling mixture, and the other is later diffused into the gel leading to a very high supersaturation of the main product and in due course to nucleation and crystal growth. This is, of course, growth from solution but without convection currents. This is because the gel medium prevents turbulence and provides a three-dimensional structure that permits the reagents to diffuse at a rate that can be controlled by varying the gel parameters. Further, since the gel is soft and isotropic, it exerts low and uniform constraining forces upon the growing crystals, promoting orderly growth. Also, the gel medium is chemically inert to most of the reactants and products. The other advantage is that the growing crystals are spatially separated, thereby preventing the interactions between them to a certain extent.

Gel Preparation

The first step in crystal growth from gels is the preparation of the gel. A gel has been defined as a two-component system of a semisolid nature rich in liquid. Closest to gel in structures are sols from which they are usually prepared. Sols, which are likewise two-component systems, resemble liquids more than solids. Further, because gels are formed by the process of gelling of sols, it is essentially the preparation of sols that needs to be carried out first. In fact, the preparation of sols is more difficult than the preparation of gels. To produce sols that are stable at a reasonably high concentration, it is necessary to grow the clusters of molecules to a certain size under alkaline conditions (clusters are negatively charged in alkaline conditions) so that they will not flocculate or gel.

Gels can be prepared from a variety of materials. The most commonly used ones for growing crystals are those prepared from agar, gelatin, silica, tetramethoxysilane (TMOS), and polyacrylamide. The procedure to form gel from these materials varies. For example, to form agar gel, 1%–2% by weight of agar-agar is dissolved in water. The clear solution is boiled and cooled down slowly. The gelation takes place during cooling [25]. Gelatin gel is prepared by dissolving gelatin in water and stirring it at a constant temperature of 50°C for an hour or so and then cooling it down to room temperature. A small quantity of formaldehyde, if added, strengthens the gel [26]. Silica gel is prepared by neutralizing the aqueous solution of sodium metasilicate with mineral or organic acids [21]. Silica gels can also be prepared by adding water to silanes, such as TMOS or tetraethoxysilane [23, and the references therein]. Polyacrylamide gel is prepared by dissolving ~4% by weight of acrylamide and 0.02% by weight of a cross-linking agent in water. The solution is bubbled with nitrogen gas and then degassed by reducing the pressure. This gives a rigid transparent gel [27]. Other gels used in crystal growth are polyethylene oxide and polyvinyl alcohol. In physical gels such as agar and gelatin the sol-to-gel transition is reversible. However, in chemical gels like silica and polyacrylamide, the transition is irreversible.

Among the various gels that can be used for growing crystals, silica gel seems to be the most widely used. Many processes have been employed for producing colloidal silica from low-cost sodium silicate solution. Dialysis, electrodialysis, ion exchange, peptization of gel, hydrolysis of silicon compounds, and neutralization of soluble silicate with acids are some of the processes used for gelation. Of these, the one that is generally employed by crystal growers is the method of neutralizing soluble silicates with acids; hence, we deal with this method in greater detail.

When a dilute solution of sodium silicate (which is alkaline) is partially neutralized with acid to a pH in the range of 8–0, a silica sol rather than a gel is obtained, provided the concentration of the resulting sodium salt is less than about 0.3 N. While making a sol by neutralizing with acids, the mixing by stirring should be carried out so rapidly that no part of the solution remains in the pH range of 5–6 for an appreciable time. This is because silicic acid would transform into gel almost instantaneously at these pH values.

Silica sol thus prepared changes into gel in due course at times that can vary widely from minutes to many days, depending on the material, its temperature, pH, and history. Initially, there is only a slow increase in viscosity. Then, the viscosity begins to increase rapidly and solidification occurs at the gel point. The gel point can easily be determined by tilting the container in which it is being set. If the gel is set, the meniscus will no longer remain horizontal when tilted. Now, depending on the density and precise conditions during gelling, the hardness of fully developed gels can vary widely from being very soft to hard. However, in most of the crystal growth studies, it is the soft gel that is used.

One of the most important factors affecting the hardness of the gel medium is the density of the sodium metasilicate solution. In almost all cases, dense and hard gels have produced poor-quality crystals. On the other hand, if the gel density is too low, it will take a very long time to set, and the resulting gel will be mechanically fragile and hence unstable. Experimenters have realized that solutions with densities in the range of 1.03–1.06 g/cc yield satisfactory results.

Silica gel can be set by adding either acid to sodium metasilicate or sodium metasilicate to the acid. The two routes are not the same because in one case the gelation pH is approached from the acid side (pH 0.5–4.0) and in the other it is from the alkaline side (pH ~12). These are sometimes called acid and alkali set gels, respectively.

Gelling Mechanism and Structure

When sodium metasilicate is dissolved in water, the following reaction would take place to produce orthosilicic acid.

$$Na_2SiO_3 + 3H_2O \equiv H_4SiO_4 + 2NaOH$$

It is generally accepted that orthosilicic acid can polymerize with the liberation of water. As mentioned earlier, the rate of polymerization depends on the pH value of the solution. At around 3–4 pH, polymerization is initiated with formation of chain-like or open branch-like structures as follows:

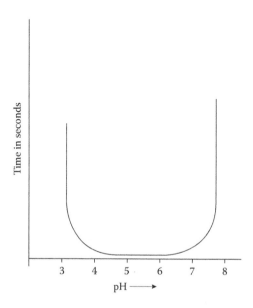

This reaction repeats all over the solution to give a three-dimensional network of —Si— O links. As the polymerization process continues, water accumulates on top of the gel surface. Expulsion of water would also occur due to gel shrinkage. Such a phenomenon is known as syneresis.

It is important to realize that the gelation time is extremely sensitive to solution pH. Figure 9.8 schematically shows the relationship between the time required for gelation and pH of the solution. During polymerization,

FIGURE 9.8
Gelation time versus pH curve.

in addition to the formation of $H_2SiO_4^-$ shown by the earlier reaction, $H_3SiO_4^-$ ions are also produced, the relative amounts of which depend on the pH value. The formation of more reactive $H_2SiO_4^-$ is favored at high pH values. Because of the higher charge of this species, there is a greater degree of repulsion between them, thus slowing down the gelation process. $H_3SiO_4^-$ is favored by moderately low pH values and is believed to be responsible for initial formation of long chains due to polymerization. Subsequently, cross-linkages are formed between these chains, leading to sharp increase in viscosity that signals the onset of gelation. At very low pH values, however, the tendency toward polymerization itself is diminished, thus slowing down the process of polymerization. This qualitatively explains the gelation time versus pH curve shown in Figure 9.8.

Some efforts have been made to study the structure of these gels by direct visualization. Transmission electron microscope study of the frozen gels gives information about the maximum size of macropores, where the nucleation and growth of the crystals occur [28]. Pore-to-pore distance has also been deduced from light-scattering experiments performed on these gels [29]. From the scanning electron microscopy study of frozen gel, it has been realized that gel actually consists of sheet-like structure of varying degrees of surface roughness and porosity, the pore sizes varying widely from 0.1 to 5 μm.

Apparatus and the Techniques

One of the overriding features of the gel method is the simplicity of the growth apparatus used. In fact, the gel may be allowed to set in a simple test tube with one of the reactants already added to it. Once the gel is set, the other reactant may be poured over the gel without damaging its surface. Alternatively, a U tube may be used with the gel set at the bottom and the reactants added to the two arms of the U tube. These two basic growth apparatus are shown in Figure 9.9. The reaction that would take place may be represented as

$$AX + BY \rightarrow AY\downarrow + BX$$

in which the desired product is insoluble and the by-product is usually highly soluble in water. This technique is termed "growth by chemical reaction."

The main drawback of the test tube method is the limited range of reagents and concentration which may be added to the gel without interfering with its setting characteristics. Also, since the quantity of the reagent that can be added is limited, the size of the crystals that can be obtained would also be limited. Although this disadvantage is overcome with the U tube method, the awkward shape of the U tube requires special support to keep it vertical. It is also more difficult to clean U tubes compared to a test tube.

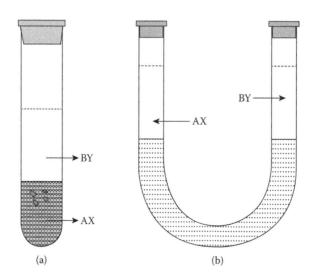

(a)   (b)

FIGURE 9.9
Schematic of gel growth in (a) a test tube and (b) a U tube.

Since the revival of the gel growth method, several improved versions of the growth apparatus have been used by crystal growers. While the basic principles involved in the experiments using these are the same as those of the U tube, they have certain advantages over the conventional apparatus. These improvisations include an open tube anchored inside another test tube, the tubes with fritted disks, U tube with large reservoirs at the top, two large reagent reservoirs connected horizontally by a straight gel column providing a linear diffusion path, and so on. Other modifications of the growth apparatus are three-reservoir systems and open-tube systems (single and double). The various variants and modifications in the growth equipment have been reviewed by Henisch [21] and later by Arora [30]. Nevertheless, test tubes and U tubes are the most widely used apparatuses in gel growth. The three-reservoir system and $CsIO_4$ crystals grown using it are shown in Figure 9.10. In this case, the metal chloride solution was poured into the central reservoir and periodic acid into the other two reservoirs [31].

There are other processes by which crystals can be grown in gels. These are growth by chemical reduction, growth by complex formation, and growth by solubility reduction. The chemical reduction method is particularly suitable for growing metallic crystals. Crystals of gold [32], lead [33], and copper [34] have been grown by this method. For example, to grow copper crystals, a suitably set gel with $CuSO_4$ as an inner reactant is taken in a test tube. Afterward, a reducing agent such as hydroxylamine hydrochloride or hypophosphoric acid is added on the top as an outer reactant. The chemical reduction of the $CuSO_4$ gives the desired copper crystal inside the gel.

(a)   (b)

FIGURE 9.10

(a) Three-reservoir system and (b) typical crystals of $CsIO_4$ grown with it. (Reprinted from *Journal of Crystal Growth*, 121, Al Dhahir, T. A., et al., Growth of alkali metal periodates from silica gel and their characterization, 132–140, 1992, with permission from Elsevier.)

The complex formation method is used to enhance the solubility of a material by adding another soluble material whose presence increases the solubility in a nonlinear way as its concentration increases. Initially, a chemical complex of the material of the crystal is formed in the solution with an appropriate substance. It is then allowed to dissociate to form the required crystal. To achieve decomplexion, the complex solution is steadily diluted while it is diffusing through the gel. Armington and O'Connor [35] have pioneered this technique for growing cuprous halide crystals.

The solubility reduction method is mainly applicable to water-soluble materials. Glocker and Soest [36] used this technique to grow ADP crystals. They diffused alcohol into a gel containing the crystal salt solution. The alcohol reduced the solubility of the compound and thereby created the nucleation, leading to the formation of the crystals. Utilizing this technique, KDP crystals of reasonable size have been grown [25].

Kinetics and Mechanism of Growth

Gels are obviously not impermeable, but they certainly suppress macroscopic convection currents. In the absence of macroscopic convection, the only mechanism available for material transport is diffusion. In fact, growth from gel media is rather a near ideal case for applying Frank's volume diffusion theory [37] based on Fick's diffusion law. Thus, one would expect the kinetics of gel-grown crystals to follow a parabolic law and consequently a linear relationship between the square of the crystal size l and duration of growth t. Figure 9.11 validates this expectation. However, there is likely to be deviation from this parabolic law during the initial transient period during which the steady-state concentrations are established as well as at the last

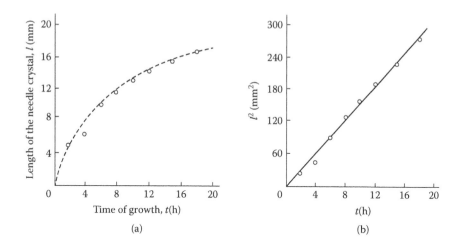

FIGURE 9.11

(a) Length of the lead chloride needle crystal versus time graph and (b) plot of l^2 vs t. (With kind permission from Springer [*Journal of Materials Science*, Filamentary and dendritic growth of lead chloride crystals in silica gel, 16, 1981, 1707–1710, Bhat, H. L.].)

stage when the available solute is exhausted. Both factors must be expected to give rise to nonlinearity.

The fact that a linear relationship between l^2 and t has been observed in both the case of growth involving chemical reaction and simple solubility reduction suggests that the chemical reaction is not the rate-determining process. It is worth noting that the parabolic relation holds true not only for three-dimensional bulky crystals but also for needle-like crystals (see Figure 9.11), which can be regarded as one-dimensional crystals [38].

Once the solute reaches the growing crystal surface by diffusion, growth takes place either by two-dimensional layer growth or by screw dislocation mechanism. Figure 9.12 shows the habit face of a lead sulfate crystal grown by the gel method with the spreading of characteristic growth layers, which appear to have initiated from the bottom edge [39].

In silica gel, the growing crystal either tears through or traps the gel as it grows. In the case of displacing the gel, cavities (regions in which the gel has been split and separated from growing faces) are formed. These cavities are filled with the solution, and neighboring cavities are interconnected. Under such a situation, the crystal is almost entirely surrounded by the solution and hence grows increasingly from solution. In the case of trapping of gel, the final crystal looks turbid and is contaminated.

Nucleation Control

As mentioned earlier, one of the major drawbacks of gel growth technique is that it gives only small crystals. Some efforts have been made to overcome this problem by reimplanting the grown crystal as a seed for further growth.

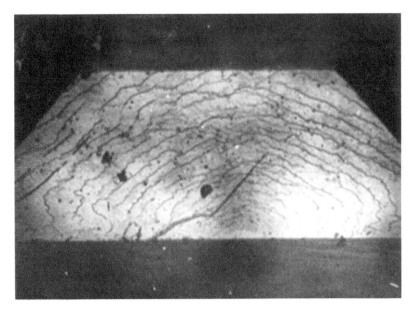

FIGURE 9.12
Growth layers observed on (101) surface of a $PbSO_4$ crystal that appear to have initiated from the bottom edge. (From Bhat, H. L., Studies on growth and defect properties of barite group crystals, PhD dissertation, Sardar Patel University, Vallabh Vidynagar, India, 1973.)

This can be done by placing the previously grown crystals on the surface of a preset gel and covering it with more sodium silicate solution, which is then allowed to set. Before adding the reagent, surface layers of the seeded crystals may be dissolved (say, by temporarily increasing the temperature) so that the surface contaminations are removed. Because the boundary between the new and old gels tends to support fresh nucleation, the old gel surface must be protected from getting contaminated. Reimplantation can also be achieved by suitably modifying the growth apparatus so that the crystals can be transferred with a small portion of gel surrounding them. The open-tube system offers such a facility.

Heterogeneous nucleation is also possible in gel growth. This has been demonstrated by deliberately mixing foreign particles in the gel and inducing the crystals to grow on them [40]. However, a large number of observations support the idea that the primary mode of nucleation in gel growth is homogeneous nucleation [21]. Therefore, the problem of limited crystal size has to be tackled first by limiting the nucleation. Other issues of relevance are ensuring continued reagent supply and removal of the waste product. Since supply problems and removal of waste products have already been discussed to some extent, here we discuss only the control of nucleation. The suppression of nucleation is the principal function of the gel, and this has to do with the pore size distribution of the gel. As expected, the smaller the size of the pores, the higher the suppression of nucleation, and this can be achieved by simply increasing the density of the gelling solution. However, this hardens the gel and the growing crystals tend to incorporate it rather than excluding it, leading to gel contamination. Therefore, the density of gel used in practice is a compromise. Another method of nucleation control involves the deliberate addition of foreign particles, and this, of course, is deliberate contamination of the growing crystals [40]. The use of a particular combination of reactants to grow the crystal has also been found to reduce the nucleation density [41]. However, there is no easy way of finding which of the combinations would yield the best result.

By far, the most useful control procedure involves concentration programming. In this procedure, initially the concentration of the diffusing reagents is kept below the level at which the nucleation is known to occur. It is then increased in a series of small steps that can be optimized for any system in terms of magnitude and interval. At some stage as the concentration of the reactants increases, a few nuclei begin to form. Since these nuclei act as sinks for the solute, they establish radial diffusion patterns around them, which actually reduce the reagent concentrations in neighboring locations. In this way, the formation of the additional nuclei would be inhibited. Subsequent increase in the reagent concentration leads to faster growth but not in general to new nucleation. It has been found empirically that frequent smaller steps are preferred to a few large concentration jumps. The method has been successfully applied to the control of nucleation in several systems [42].

Hydrothermal Growth

The growth of single crystals from aqueous solution is usually performed by a careful control of supersaturation of the solution in the presence of a seed crystal. The process is generally carried out at atmospheric pressure. However, if the normal solution growth is unsuccessful because of low solubility, one can still try solution growth by suitably modifying the growth conditions, thereby increasing the solvent action, or using other solvents with greater solvent powers. One way of increasing the solvent action is to use a mineralizer, which forms additional species (complexes) that increase the overall solubility. The addition of mineralizers can be used only when the complexes formed with them are not stable solids under the prevailing growth conditions. One can also increase the solubility by conducting the growth at elevated temperatures. However, in the hydrothermal growth, an aqueous solution is held not only at high temperature but also at high pressure so that the nutrient that is otherwise insoluble at ambient conditions dissolves appreciably. Additionally, a mineralizer is also used to enhance the solubility. Some of the mineralizers used in hydrothermal growth are NaOH, KOH, NH_4Cl, NaCl, KCl, KF, CsF, Na_2CO_3, and so on.

Hydrothermal growth is normally carried out in a sealed vertical auto-clave with a temperature gradient established between the top and bottom of the vessel. Usually, the nutrient is placed in the lower, hotter part of the autoclave, and the seeds are mounted in the cooler, upper part. The seeds are single crystal plates properly oriented and mounted on a suitable wire frame. A perforated metal disc called the baffle is often placed to separate the dissolution and growth regions so that the temperature gradient is local-ized. A schematic of the arrangement is shown in Figure 9.13. Transport of hot solution from the dissolution region to the growth region containing the seeds is by convection. Once the solution reaches the growth region, it becomes supersaturated with respect to the temperature of that region and the material is deposited on the seeds. The cooler depleted solution then returns to the hotter zone by convection and dissolves more nutrients, and the cycle repeats.

Because hydrothermal growth is quite involved in terms of both equipment and procedure, not many laboratories practice this technique. However, in recent years, low-pressure and moderate-temperature hydrothermal synthe-sis has gained prominence particularly in the preparation of nanoparticles. Excellent reviews on these aspects are available in the literature [43,44]. Here, we discuss only the conventional hydrothermal growth. In writing this sec-tion on hydrothermal growth, we have relied on the classic review written by Laudise and Nielsen [45].

Autoclaves

The success of the hydrothermal crystal growth experiments depends mainly on the quality of the autoclaves, which contain corrosive aqueous solution at high temperature and pressure. Therefore, the first consideration in the fabrication of an autoclave is the material of construction. For tempera-ture in the range of 250°C–300°C, sealed glass or vitreous silica tubes have been used fairly successfully up to 6 atm. The use of glass in hydrothermal growth has not been explored fully because early works were mainly on the growth of materials that could not be dissolved readily in the pressure–temperature (P–T) range attainable in glass autoclaves. However, it has been realized that there is a wide range of materials that can be crystallized in the range of temperature and pressure attainable in glass. Further, glass is not corroded by neutral and acidic solutions (except hydro fluoric acid (HF)). Being transparent, it allows constant observation and measurement during growth. For pressures above 6 atm, it is necessary to use steel or its alloys.

Many designs have been developed for the fabrication of autoclaves. Different designs have different pressure-holding capabilities. Proper selec-tion of the autoclave is necessary to prevent failures. The autoclave failures may either be catastrophic or noncatastrophic. The catastrophic failure is

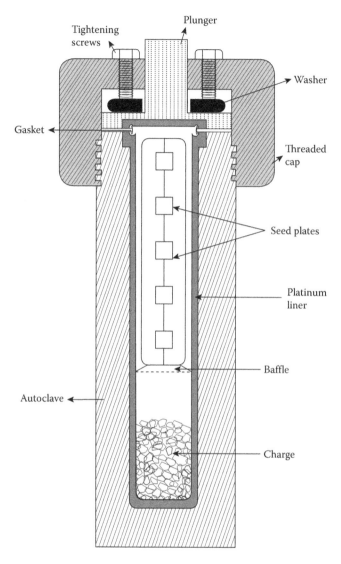

FIGURE 9.13
Schematic of hydrothermal growth.

usually associated with brittle fracture and the noncatastrophic to the failure of the pressure seal. When appropriate materials are chosen and care is taken during the operation, autoclaves can be repeatedly used for as many as 10 years. The types of autoclaves currently being used are flat plate closure autoclave due to Morey and Niggli [46], cold seal cone closure by Roy and Tuttle [47], welded closure by Walker and Buehler [48], and Bridgman (both full Bridgman and modified Bridgman) [49,50]. Of these, closure developed by Bridgman and modified by several commercial vendors is perhaps the most useful design for hydrothermal growth.

Full Bridgman

The original design is shown in Figure 9.14a [49,50]. Here, the region A of the vessel B where the nutrient is present is under pressure. The piston C is free to rise upward until stopped by the main nut E, which is threaded to B. D is a deformable gasket. For hydrothermal systems, steel gaskets are almost always required. The clearances must be reduced by the back washers F since gasket material is expected to undergo only elastic deformation. The initial seal is caused by mechanically tightening the plunger against the gasket by means of the set screws G shown in Figure 9.14a. The high-pressure seal is caused when the force of the pressure generated due to heating acts on the piston moving it upward against the gasket. One of the difficulties of the full Bridgman closure is the tendency for spontaneous nucleation between the piston surface and the wall during growth. When spontaneous nucleation occurs, the threads may get jammed so tightly that one may have to destroy the autoclave to open it.

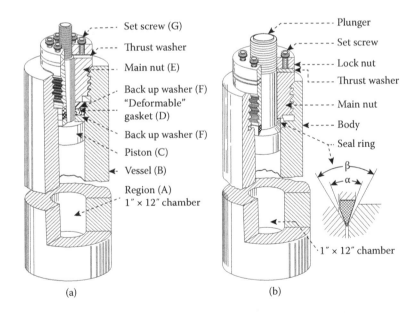

(a)   (b)

FIGURE 9.14
(a) Full Bridgman autoclave and (b) modified Bridgman autoclave. (Panel a is from Bridgman, P. W., *Proceedings of the American Academy of Arts and Sciences*, 49, 627–643, 1914. With permission. Panel b is from Ballman, A. A., and Laudise, R. A., *The art and science of growing crystals*, John Wiley, New York, 1963, pp. 231–251. With permission.)

Modified Bridgman

The difficulty of jamming is effectively overcome by the modified Bridgman closure shown in Figure 9.14b. In the modified Bridgman seal, the wedge-shaped seal ring initially makes line contact with the body and the plunger. However, since the ratio (area of the plunger/area of the line contact) in the seal is large, a slight internal pressure transmitted through the plunger will cause the seal ring to deform elastically to produce a surface contact. Even after deformation, this ratio is still large enough to make the seal self-energized. In this design, there are no free surfaces available where spontaneous nucleation can occur, thereby avoiding jamming of the closure [50].

The Bridgman autoclaves can be used quite successfully up to 500°C and 3700 atm pressure atm range when suitable construction materials are chosen to meet the required experimental conditions. Many modified versions of Bridgman closures have been developed and marketed by companies over the years. The details of some of them are available in the literature [43–45].

Except for a low-pressure glass system, in all other autoclaves noble metal liners or plating is used to prevent corrosion of the vessel walls. In fact, platinum has been found to be the most suitable lining material since it is inert under most conditions. In large-diameter vessels, Pt or Ag cans have been used quite successfully. If the can and the space between the cans and the pressure vessels are filled to the same percentage of their free volume, in principle the metal can will be under no pressure. This pressure-balance technique permits the use of thin-walled cans, thereby significantly reducing the cost. If the solvent is not too corrosive, both the spaces could be filled with the solvent; otherwise, the outer space is usually filled with water. The can may be slightly overfilled so that if any leak occurs, it will be from the can to the space between the can and the vessel. This will ensure the purity of the growing crystal.

Vessel threads should be coated with suitable lubricants such as graphite in an oil or water medium and suspended molybdenum sulfide or copper lubricants. While graphite-based lubricants are suitable for runs below 400°C, the other lubricants are used above 400°C.

Heating and Temperature Control

The furnace for hydrothermal growth depends on the degree of temperature control and gradient to be used for crystallization. For moderate temperatures (say, below 400°C), even hot plates with nichrome heating coil are adequate. For higher temperatures, cylindrical wire (Nichrome or Kanthal) wound furnaces with ceramic muffle are used. The spacing between the windings may be varied to produce a variety of temperature gradients. Alternatively, multizone furnaces can be used, which allow flexibility in power inputs, so that any desired temperature differential may be achieved. A variety of temperature controllers starting from simple "on–off" type to highly sophisticated ones are used.

Pressure Measurements

In most of the hydrothermal growth experiments, the pressure is estimated from the degree of filling of the free volume in the vessel and the temperature of operation. Since the solutions are largely water based in hydrothermal growth, one should expect the pressure to be close to that of vapor pressure of water at the autoclave temperature. Hence, if the percentage fill and the temperature of operation are known, one can directly find out the approximate pressure using PVT diagram of water. Kennedy has given pressure volume temperature (PVT) data for pure water in an isothermal system, and it is reproduced in Figure 9.15 for various percentages of filling [51]. This 60-year-old graph is still quite popular and is routinely used by crystal growers around the world. As can be seen from Figure 9.15, for a low degree of fill, the pressure follows the chain-dotted line up to the critical point of water. When water is replaced by 1 M NaOH solution saturated with quartz, the P–T diagram still looks similar to the one shown in Figure 9.15. Data obtained by Kolb et al. [52] indicate that the depression in pressure from a pure water system is only a few percent.

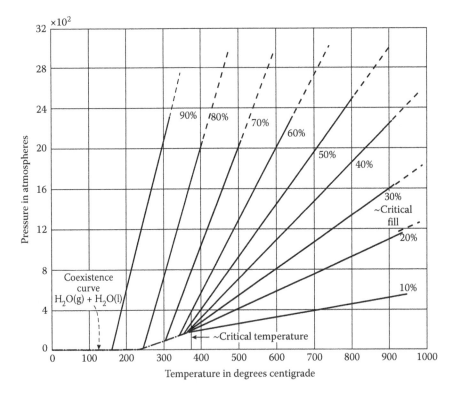

FIGURE 9.15
Pressure–temperature diagram for water at various percentage fill. (From Ballman, A. A., and Laudise, R. A., *The art and science of growing crystals*, John Wiley, New York, NY, pp. 231–251, 1963. With permission.)

Hydrothermal Growth of Quartz

Quartz is probably the most extensively studied crystal in terms of hydro-thermal growth. For quartz, the hydrothermal technique has been found to be the most suitable growth method (or the only viable method). The first successful attempt to grow this crystal by the hydrothermal method was by Spesia way back in 1905 [53]. Subsequently, many scientists have grown this crystal. Since the α–β phase transition of quartz occurs at 573°C, α-quartz, the piezoelectric modification has to be grown below this temperature. However, under hydrothermal conditions silica does not have enough solubility in pure water to permit crystallization below 573°C, but in the presence of a mineralizer such as NaOH or Na_2CO_3 it has reasonable solubility. At 80%–85% fill, the log of solubility in 0.5 M NaOH has been found to be a linear function of $1/T°K$ [54]. Although solubility of quartz increases with NaOH concentration, one cannot increase the mineralizer concentration indefinitely because it may get into the growing crystal as an impurity. For the growth, 1 M NaOH solution has been found to be opti-mum. Mixed solvents such as (NaCl + KCl) and ($NaOH + Na_2CO_3$) have also been used in recent times to grow the crystal [55,56].

Typical operating parameters for growing quartz are as follows:

Dissolution temperature	400°C
Crystallization temperature	360°C
Temperature gradient	40°C
Percentage fill	80%
Solution	1 M NaOH
Baffle open	5%
Pressure	1500 atm
Seed orientation	(0001)

Because the addition of lithium is known to improve the growth rate, a small amount of some lithium salt is usually added. Figure 9.16 shows a syn-thetic quartz crystal grown by the hydrothermal technique.

It has been observed that quartz crystal has a tendency to either reject the impurities or not grow at all in the presence of impurities, implying that its structure is not very receptive to impurities. As a matter of fact, quartz crys-tal available in nature is one of the purest minerals known and hence used in many devices. However, it is possible to grow quartz crystals with small amounts of certain impurities. Some of the impurities that can be incorpo-rated into synthetic quartz are listed in Table 9.2.

Further, it has been found that the concentration of impurities incorpo-rated in quartz depends on the crystallographic orientation of the seed on which the growth occurs.

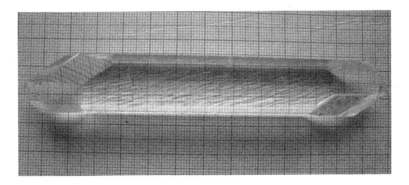

FIGURE 9.16
A quartz crystal grown by hydrothermal technique.

TABLE 9.2

Impurities That Can Be Incorporated into Quartz Crystal

Substitution for Silicon	Interstitial in Channels
Ga^{3+}, Fe^{3+}, B^{3+}, Al^{3+}	H^+, Li^+, Na^+, K^+, Rb^+, Ag^+
Ti^{4+}, Ge^{4+}	Pb^{2+}, Be^{2+}, Ca^{2+}, Mg^{2+}, Fe^{2+}, Sr^{2+}

Growth of Other Materials

The hydrothermal technique has been applied to grow many other crystals. In attempting to grow these crystals, workers have heavily relied on the abundance of information available on the quartz system. The thermodynamic and kinetic considerations applicable to the quartz system can be extended to many other systems also. Nevertheless, the operating conditions are material specific, and one has to determine these either empirically or through painstaking investigation. A few representative crystals that have been grown by the hydrothermal method are listed in Table 9.3 along with typical growth parameters. The references given here are not necessarily the first report on the hydrothermal growth of the selected material. Exhaustive tables can be found in Ref. [44].

Merits and Demerits

Even though hydrothermal growth is labeled as a high-temperature process, the crystallization temperatures here are rather low compared to the melting point of the material being grown. Consequently, the crystal experiences fewer thermal stresses and therefore is likely to have lower defect concentration than that grown from its melt. The low growth temperature also permits the growth of a low-temperature phase of

TABLE 9.3

Representative Crystals Grown by Hydrothermal Method with Typical Growth Parameters

Material	Max. and Min. Temperatures (°C)	Pressure (kbar)	Mineralizer	Reference
α-Al_2O_3 (sapphire)	620–300	0.5–0.7	NaOH, KOH, Na_2CO_3, K_2CO_3	[57]
$Y_3Fe_5O_{12}$ (YIG)	420	2.0	KOH	[58]
$Y_3Al_5O_{12}$ (YAG)	370	2.0	NaOH, Na_2CO_3, K_2CO_3	[59]
$CaCO_3$ (calcite)	450–200	0.6–0.8	NH_4Cl, LiCl, H_2CO_3, HNO_3	[60]
Al PO_4 (berlinite)	300–150	0.03–0.2	H_3PO_4, H_2SO_4, HCl, HCOOH	[61]
$Li_2B_4O_7$ (LBO)			HCl	[62]
$KTiOPO_4$ (KTP)	600–425	10 kpsi	K_2HPO_4 + KPO_3	[63]

certain materials (e.g., quartz), which are often not possible by melt-based techniques. Because of the lower viscosity of hydrothermal solutions, there is rapid convection leading to efficient material transport, resulting in a relatively rapid growth rate for a solvent–solute system.

The disadvantages of the technique are in the complexity of the method, the high cost involved in the design, the fabrication of the high-pressure autoclaves, and the need for corrosion-resistant autoclave liners. Also, because the autoclaves are closed systems, the growth as such cannot be observed (except for glass autoclaves). Thus, the success or otherwise of a growth run can be ascertained only after the entire run is over.

However, the key limitation of the conventional hydrothermal growth is in the process development, which is still largely based on time-consuming, empirical, trial-and-error methods. The development of a rational approach is yet to take place, and the current research is focused on this. This is being attempted through studies on phase equilibria to generate the equilibrium phase diagrams, computational thermodynamics, experimental validation of computational results, and so on.

Flux Growth

In growth from solution, which was discussed earlier in the section "Growth from Aqueous Solution" of this chapter, the solvent is always a liquid around room temperature. The solution growth methods have been extended by the use of molten salt solvents, the technique being called flux growth.

Here, the crystallization is carried out using molten inorganic compounds at elevated temperatures as a solvent and hence is actually a high-temperature solution growth process. It is also sometimes referred to as molten salt solvent growth technique. The solvents are usually called fluxes. Because there are innumerable chemical compounds available that can be used as fluxes for growing crystals, it is probably true that any material can be grown by this technique. Consequently, it is regarded as the most versatile polycomponent growth technique. Even though the technique has been known since the late nineteenth century, full exploitation of the technique began only after World War II. This was mainly due to the widespread demand for a variety of high-melting-point inorganic crystals with unusual properties that could be used to make solid-state devices. $BaTiO_3$, a room temperature ferroelectric, is one such crystal that drew the early attention of crystal growers. Because of the structural transition in $BaTiO_3$ below its melting point, its growth from pure melt was not possible. Even though there were earlier attempts, Remeika [64] grew large crystal plates of $BaTiO_3$ in the characteristic butterfly habit from KF solution. Since then, the method has been extensively used to grow hundreds of crystals of scientific and technological importance.

Literature available on flux growth is extensive. *Crystal Growth from High Temperature Solution*, written by Elwell and Scheel [65], is mainly devoted to flux growth and is the most authoritative book on the subject covering the work up to 1975. It also traces the history of crystal growth from high-temperature solutions. Chapters written in the books *The Art and Science of Growing Crystals* [66], *Techniques of Inorganic Chemistry*, Vol. IV [67], *Handbook of Crystal Growth*, Vol. 2 [68], and *Springer Handbook of Crystal Growth* [69] all give useful information on flux growth, the last two dealing with the later developments.

Principle

The principles of flux technique can be explained with the help of a simple solvent–solute phase diagram with a eutectic shown in Figure 9.17. Referring to the figure, the solution of composition n_A equilibrated at a temperature T_A (point A) may be slowly cooled without any disturbance to a temperature T_B (point B). Obviously, the region between the liquidus and the line passing through the point B is the metastable region, which is undercooled or supersaturated. At this point, nucleation occurs and the growth starts. Continued slow cooling from T_B results in growth on already nucleated crystals and, as indicated by the crystallization path in Figure 9.17, with a much lower level of supersaturation. On the other hand, if the growth is by solvent evaporation at constant temperature, the path traced is from A to F at which point the nucleation occurs. If the solution is transported from a hotter region to a cooler region to achieve supersaturation and subsequent nucleation, one obviously follows the path C to D. All three methods are used in flux growth.

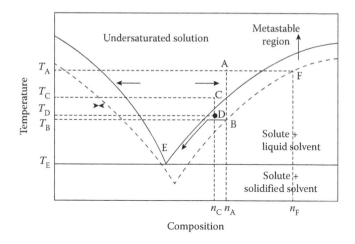

FIGURE 9.17
Typical solvent–solute phase diagram with a eutectic.

Solvents and Solubility

The selection of a suitable solvent for the flux growth is perhaps the greatest challenge the crystal grower faces. An ideal solvent should have the following characteristics:

1. The solute should be the only stable solid phase under the growth conditions.
2. The solute should have a solubility of about 10%–50% in the solvent.
3. The solute should have an appreciable temperature coefficient of solubility (say, ~1 wt% change for 10°C change).
4. The solvent should have low volatility.
5. The solvent should not react with the crucible material.
6. Solvent solubility in the grown crystal should be very low.
7. It should have low viscosity to facilitate effective mass transport.
8. The solvent should have a low melting point.
9. It should have well-separated (large difference) melting and boiling points.
10. The solvent should be easily leachable.

In practice, no solvent is completely satisfactory in terms of all the aforementioned requirements, and some compromise has to be made. Solvent selection is usually guided by intuition and analogy with the known systems. Usually, materials that meet the most important properties would be chosen as the solvent. A good compilation of phase equilibria data [70,71] is helpful in selecting a proper solvent. Also certain modification in experimental conditions can help in relaxing certain requirements. For example, covered or welded crucibles would relax the requirement of low volatility.

Certain chemical considerations can help in the selection of solvents. For instance, acidic oxides are good solvents for basic crystals and vice versa, provided the salts are not stable solid phases under the growth conditions. Also, complex formers will be useful solvents, provided complexes are not solids. While chain-breaking cations are useful in lowering the melt viscosity, nonsimilarity between the solute and solvent would help in suppressing solid solubility. Further, a common ion between the solute and solvent is helpful to suppress the solvent contamination. However, what is important for a successful growth is the knowledge of solubility and its temperature coefficient. Unfortunately, it would be too lengthy a process to determine the solubility curves for many different solvent–solute systems. For a preliminary screening, it is often more convenient to carry out a few small-scale experiments with different solvents using a solute–solvent molar ratio of about 1:10. The products formed on slow cooling will often give a clear indication of either crystallization or chemical reaction. Those showing reasonably large crystals (size ~1 mm) can then be investigated in more detail.

Lead compounds have been found particularly useful as solvents, and varieties of crystals have been grown from solutions of lead oxide, lead fluoride, lead borate, or lead pyro phosphate. The bonding characteristics of Pb^{2+} coupled with the large ionic size are suggested to be the contributing factors to this. Since lead compounds are highly toxic, extreme precautions must be taken in handling them. Apart from lead-based compounds, many different materials have been used as flux for the temperature range (~700°C–1400°C) in which most flux growth experiments are performed. A list of a large number of solvents has been given by Elwell and Scheel in their book [65]. They range from simple compounds to a mixture of variable compositions. The summary of fluxes used by Wanklyn and coworkers in growing various magnetic oxides is quite educative [72].

Determination of solubility at high temperature is not that easy. The solubility at a given temperature can be determined approximately by slowly cooling a number of small crucibles containing different solvent–solute ratios. If a marked increase in crystal size is observed for a particular composition, it indicates that the liquidus temperature is exceeded for that composition. If loss of solvent by evaporation is high, small sealed platinum tubes may be used. These are filled with weighed amounts of solvent and solute fragments. After heating, the tubes are quenched and weighed to ensure that no loss of solvent has occurred. The solubility is then determined by weighing the undissolved solute fragments and estimating the loss of weight of the solute fragments.

Apparatus

The apparatus required for flux growth is rather simple. First, a chemically nonreactive container capable of withstanding high temperature is required. The usual temperatures involved in a flux growth range from about 700°C to 1400°C; hence, a well-insulated furnace capable of reaching this temperature is the second requirement. Since the cooling protocols are material specific, a versatile temperature programmer and controller is the third requirement. We discuss these here in some detail.

Crucibles

For flux growth, platinum has been found to be the best material because of its ability to withstand solvent attack as well as high temperature. Consequently, we see its extensive use in this technique. Even though the initial cost of platinum ware is high, it has the advantage of being used again and again. Also, it can be reshaped and used. Further, the recovery value for scrap platinum is high, and replacement of crucible may cost only about 10%–15% of the initial price. Alloys of platinum with rhodium and iridium, which confer additional robustness and increased yield strength, have also been used. However, these alloys are more susceptible to solvent attacks. While using lead compounds as flux in platinum crucible, one has to be careful because these compounds have a tendency to get reduced to form metallic lead. Since metallic lead forms a low-melting alloy with platinum, it often leads to crucible leakage and failure. Therefore, it is important to avoid reducing conditions when platinum crucibles contain lead compounds as solvents. Bi^{3+} compounds also behave similarly to lead compounds.

Crucible geometry is dictated by the furnace configuration and the desired temperature profile. While standard geometry is adequate for small-scale experiments (100–200 cc), cylindrical geometry is preferred when crucibles are large (i.e., greater than a liter). Evaporation of the solvent is prevented by closing the crucible with the lid tightly by crimping. Sometimes the lid is welded to the crucible flange for a better seal. Soldering may also be performed with Pd wire as the solder (melting point of Pt = 1773°C and melting point of Pd = 1534°C). When a seal is caused by welding, one normally drills a small hole in the covering lid to allow the moisture and entrapped gases to escape. Once the charge has become dry, the hole may be plugged by a platinum wire. On completion of the experiment, the welded joint is carefully cut away, leaving sufficient projection for subsequent welding.

Furnaces

Even though preliminary experiments can be performed in small furnaces, actual flux growth experiments require well-insulated large furnaces where in a substantial volume can be heated for a considerable length of time.

These can be tubular or box type. Several types of electrical furnaces have been used for the purpose, including conventional wire-wound muffle furnaces. Muffles with high thermal conductivity improve the thermal homogeneity. For temperatures above 1200°C, SiC rod furnaces are more suitable because one can work with SiC rod furnaces safely up to 1350°C for an extended period. However, for maximizing the life of SiC heating elements, a temperature of 1400°C should not be exceeded for long periods. Since the resistance of SiC heating elements is known to change with use, it is necessary to adjust the input voltage with time. Temperatures are generally measured by Pt–Pt 10% Rh thermocouples.

Temperature Control

Temperature control is one of the most important aspects of flux growth. In the case of growth by slow cooling, it is necessary to reduce the temperature according to a predetermined program. Preliminary runs can be made with the help of "on–off" or proportional controllers. However, for actual growth runs, sophisticated programmable temperature controllers are always preferred. Eurotherm programmable temperature controllers are widely used by the crystal growth community for this purpose.

Stirring

The viscosities encountered in flux growth systems are high compared to those in room temperature solution growth and hydrothermal growth. To achieve uniform growth and minimize the tendency for the inclusion of the solvent in the growing crystal, it is necessary to rotate the crystal or to stir the solution. In some instances, both crucible and seed are rotated. For effective mixing, the accelerated crucible rotation technique (ACRT) has been employed by many workers. This method, for example, has been applied to the growth of $Y_3Fe_5O_{12}$ in a spherical crucible [73]. In case of volatile solvents, one has to worry about the design of a suitable rotating seal that allows for rotation and yet minimizes the loss of solvent by evaporation. Several other stirring schemes are also reported in the literature.

Techniques of Growth

Since flux growth is nothing but solution growth at high temperatures, many similarities exist between procedures followed in flux growth and those adopted in growth from aqueous solution. Also, the basic principles of growth are the same for both methods. For example, in flux growth, crystallization may occur either by spontaneous nucleation or heterogeneous nucleation. Growth can also occur by the controlled nucleation on seed crystals. Further, as in solution growth, the supersaturation may be achieved by (1) slow cooling, (2) solvent evaporation, and (3) thermal gradient. In early experiments, it is likely that a combination of these conditions was employed. However, with the advancement in technology, methods to isolate them are now available.

Growth by Slow Cooling

In this method, the mixture of solute and the solvent, which are both solids at room temperature, taken in the required proportion is filled in an appropriate crucible. The crucible is then placed in the isothermal region of the furnace with the lid tightly covering the crucible. The temperature of the furnace is raised to the required level and held there for some time for homogenization (soaking). Subsequently, the cooling is commenced in a programmed manner. The solution should preferably become saturated at some 20°C–30°C below the homogenization temperature. Figure 9.18 shows the schematic of a typical flux growth arrangement.

Ideally, constant supersaturation should be maintained throughout the growth, which requires the cooling rate to be increased progressively. This is because the area of crystal surface available for growth increases as the growth proceeds. However, in practice, often the cooling rate is kept approximately constant. Slow cooling is maintained until the solidification of the residual melt begins. At this stage, one can decant the residual solution from the crucible, leaving the crystals attached to the base or walls of the crucible. This procedure is followed particularly in cases where the solvent is relatively insoluble in common leaching agents. Alternately, the crucible is cooled to room temperature, and the crystals are extracted by leaching away the flux using a suitable leaching agent.

In flux growth, crystals usually nucleate first in the coolest region of the crucible. Even where great effort has been made to keep the crucible isothermal, some regions remain slightly cooler than the rest and the nucleation begins. Nucleation in flux growth is usually heterogeneous at the walls of the crucible, although spontaneous nucleation cannot be completely ruled out. If a small gradient is established between the top and bottom of the crucible, with the bottom being at a lower temperature, crystals grow at the bottom or lower side wall of the crucible. This variant is sometimes referred to as slow-cooling bottom growth and was adopted by Tolksdorf [73] to grow YIG crystals in a spherical crucible. In his experiment, the crystal seed is fixed in the upper part of the spherical crucible; the melt is homogenized at 1250°C for 1 day and supersaturated by cooling to 1075°C. The crystal seed is now rotated downward to bring it into contact with the solution. An air stream of 170 L/h cools the seed by about 20°C. Growth is performed with ACRT between 1075°C and 1000°C by cooling at 0.2°C/h during 175 h and then 0.5°C/h during 80 h. Finally, the crystal is rotated upward in order to avoid the adhesion of the flux at the crystal surface, and the system is cooled at 60°C/h to room temperature. Figure 9.19 [73] shows a YIG crystal grown from 3 kg of melt.

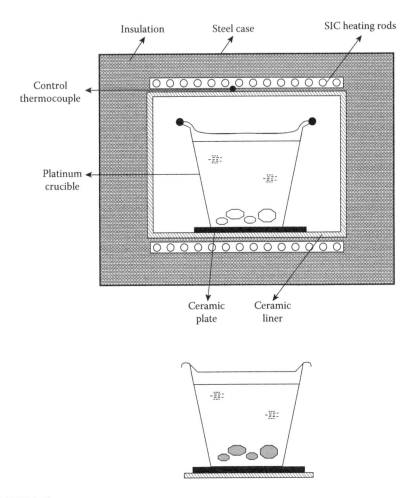

FIGURE 9.18
Schematic of a typical flux growth arrangement. Note that the crucible lid is welded to the rim of the crucible. Crucible with loosely covered lid is shown alongside.

As in aqueous solution growth, the size of the crystals obtained for a given growth protocol increases with the volume of the melt used. This is essentially because of the greater temperature stability provided by the larger thermal mass. Therefore, it is always advantageous to fill the crucible to its capacity with compacted charge.

Growth by Evaporation

Supersaturation of the molten solution can also be achieved by the slow evaporation of the flux [47]. The method is analogous to that used in solution growth. The crucible with its content is held at a constant temperature.

FIGURE 9.19
YIG crystals weighing 250 g grown from a 3 kg melt. (Reprinted from *Journal of Crystal Growth*, 42, Tolksdorf, W., Growth of magnetic garnet single crystals from high temperature solution, 275–283, 1977.)

As the flux evaporates, the solution becomes supersaturated and the crystals start developing. The control of evaporation may be achieved either by properly choosing the growth temperature or by limiting the crucible opening with a baffle. Typical $BaSO_4$ crystals grown by flux evaporation technique are shown in Figure 9.20 [74]. Here, the growth was carried out in a 25 cc crucible with a loosely covered lid. Because the flux was NaCl, it was allowed to escape and condense on the cooler parts of the muffle. However, in the case of toxic solvents, suitable arrangements have to be made to condense the solvent so that it does not escape to the surroundings. Several schemes have been invented and are described in Ref. 65.

The flux evaporation method has certain distinct advantages. First of all, it is an isothermal growth; hence, the entire crystal grows more or less at a constant temperature. Consequently, crystals will contain a lower concentration of thermally related defects. Also, the temperature control is easy. The technique is particularly useful for systems where the crystallization temperature is restricted. For example, in the growth of MgO using MgO–PbF_2 systems, MgF_2 starts crystallizing below 1150°C [75], which restricts the lower temperature to which the system can be cooled, leading to insufficient yield.

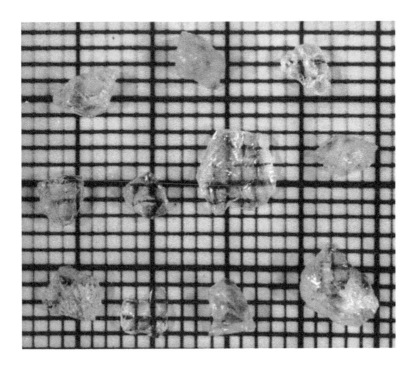

FIGURE 9.20
Typical BaSO$_4$ crystal grown by flux evaporation method. (Reprinted from *Journal of Crystal Growth*, 11, Patel, A. R., and Bhat, H. L., Growth of barite group crystals by the flux evaporation method, 166–170, 1971, with permission from Elsevier.)

Crystal growth by thermal gradient is performed in a specially designed crucible and invariably on a seed crystal. Details of this method are discussed in the following section.

Nucleation Control

In flux growth, since the growth occurs on randomly nucleated crystals, the result will be a large number of small crystals. From a practical point of view, it is preferable that the nucleation be confined to as few initial sites as possible so that larger crystals can be obtained. Optimizing the cooling rate such that no additional nuclei are formed has been found useful. The crucible material as well as the undissolved particles of the solute in the solution can provide sites for nucleation. Therefore, it is desirable for the crucible to have clean and polished inner surfaces, and the charges become a completely homogeneous solution in the initial heating. For this reason, the charge is soaked at the maximum temperature for some time before cooling is commenced. The duration of soaking varies from system to system and may be between 2 and 24 h. For highly viscous solutions, it can be much longer. Thermal cycling around the saturation temperature also helps in nucleation control [76,77]. In this process, the temperature is taken up and down a few degrees (2°C–20°C) several times around the saturation temperature with a cycling period of a few minutes, which dissolves most of the smaller nuclei, leaving only the larger ones to grow further. Another approach to reduce the number of nucleated crystals is to add a small quantity of some additives, which do not get into the growing crystal. Such additives are supposed to increase the metastable zone width, probably due to complex formation. The most widely used additive is B_2O_3. The other complex-forming agent used is V_2O_5. Certain divalent metal oxides have also been used as complex-forming agents.

The best control over nucleation is achieved, however, when growth occurs on a suitable seed crystal. This will, of course, present some difficulty in operation because the survival of the seed crystal introduced at the beginning of a growth run cannot be guaranteed without an accurate knowledge of the solubility characteristics of the systems and an accurate temperature-controlling mechanism. In spite of these difficulties, seeded growth is an attractive improvement over the growth by random nucleation in terms of control over orientation of the growing crystal, impurity incorporation, and growth rate. Two classes of seeded growth experiments are in vogue. They are (1) growth on a seed by thermal gradient and (2) top-seeded solution growth.

Growth by Thermal Gradient

As mentioned earlier, the growth by thermal gradient is performed in a specially designed crucible. The schematic of one such crucible is shown in Figure 9.21. The crucible may be considered as having two zones. The bottom hot zone contains the source material. Here, the heat introduced causes the solute to dissolve in the solvent to form a saturated solution. The lighter hot solution tends to rise to the top zone. The material transport that occurs by convection is effectively controlled by the opening of the baffle, which separates the hot and cool zone, as in hydrothermal growth. The solute transport can also be controlled by a suitable constriction of the crucible at the middle. The top cool zone contains the seeds. The rising solution gets supersaturated here, and the material deposits on the seed. The depleted solution again comes down and dissolves more material, and the cycle repeats. We just saw earlier another complicated version of a seeding technique employed to grow large YAG crystals.

Top-Seeded Solution Growth

In recent times, the most popular means of nucleation control has been to employ the top-seeded solution growth (TSSG) method. The name "top-seeded solution growth" was introduced by Linz et al. [78] in 1965. The method cleverly combines both Czochralski and conventional flux growth techniques.

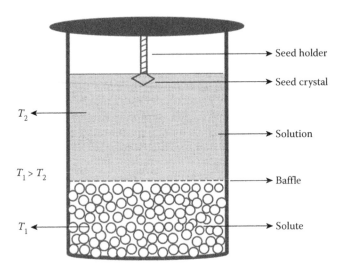

FIGURE 9.21
Typical arrangement for growth by thermal gradient method. The seed is at the top cooler part
of the crucible.

However, there is a fundamental difference between the two methods.
In TSSG, the liquid–solid interface is below the liquid level, whereas in the
Czochralski method it is generally just above the melt level. Consequently,
the temperature gradient at the liquid–solid interface is less steep in TSSG,
and this promotes facet formation which is normally observed in TSSG-grown
crystals. In a typical growth, the charge (solute + solvent) is melted in a crucible whose top is slightly at a lower temperature from the bottom. On homogenization, a seed is inserted from the top so that it makes contact with the
solution and allowed to grow. At the end of the growth, it is pulled out of the
solution and slowly cooled down to room temperature.

As an example, we describe the growth of cesium lithium borate (CLBO),
an nonlinear optical (NLO) material of technological importance, in some
detail. CLBO ($CsLiB_6O_{10}$) was first developed by Sasaki et al. [79]. Figure 9.22
schematically shows a TSSG setup fabricated in the author's laboratory to
grow CLBO crystals along with its photograph [80]. It consists of the mainframe designed to hold the growth station (three-zone furnace with crucible
holder), crucible and seed rotation mechanism, translation unit, a charge-
coupled device (CCD), and a monitor to aid the seeding.

The capability of the setup is demonstrated by growing high-quality crystals of CLBO, a borate compound, which is known to have a viscosity of
10–250 P at around its melting point depending on the composition of the
charge. The lowest value of viscosity corresponds to boron-deficient composition $Cs_2O:Li_2O:B_2O_3::1:1:5.5$; hence, this composition was used for the crystal
growth. Even the lower limit of 10 P is quite high compared to 19 cP for the
well-known non-borate crystal such as lithium niobate near its melting point.

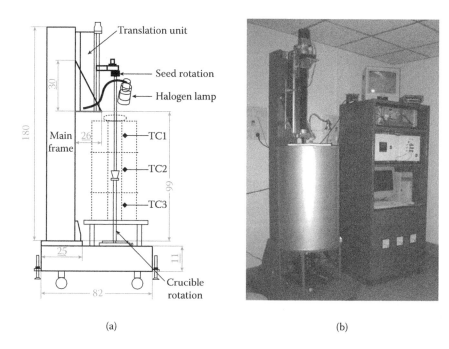

(a)   (b)

FIGURE 9.22
(a) Schematic of the TSSG mainframe (dimensions in centimeter) and (b) photograph of the TSSG unit. (Reprinted with permission from [Babu Reddy, J. N., et al., Development of a versatile high temperature top seeded solution growth unit for growing cesium lithium borate crystals, *Review of Scientific Instruments* 80, 013908, 2009], American Institute of Physics.)

To prepare the charge, the starting materials Li_2CO_3, Cs_2CO_3, and B_2O_3 taken in the aforementioned proportion are dissolved in triple-distilled water to form a clear solution. The solution is heated to evaporate water. This method of synthesis ensures better homogenization compared to mixing the ingredients in solid state. The powder so obtained is sintered at 800°C for 24 h to complete the compound formation. Thus, the formed compound is filled in a platinum crucible and heated to 900°C at the rate of 100°C/h and held at that temperature for 3 h. It is then cooled to the saturation temperature (~845°C) at the rate of 20°C/h, at which point an oriented seed is brought in contact with the melt. The crystal growth is driven by the diffusion of the material toward the growth interface, and the high viscosity of the melt hinders the material transport. To enhance the material transport, both crucible and the seed are rotated, which is possible in the growth unit described here. The seed is rotated at a maximum rate of 20 rpm, applying the ACRT. The direction of the seed rotation is reversed every 5 minutes. During the growth, the cooling rate is maintained at 0.01°C/h. From the designed setup, CLBO crystals of high optical quality measuring $50 \times 40 \times 40$ mm³ could be grown (Figure 9.23) [80].

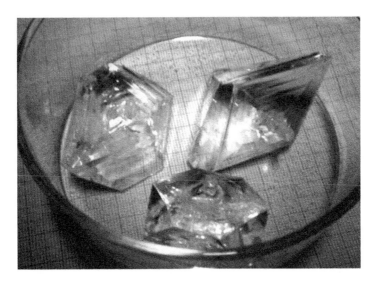

FIGURE 9.23
Typical CLBO crystals grown by the TSSG technique. (Reprinted with permission from [Babu Reddy, J. N., et al., Development of a versatile high temperature top seeded solution growth unit for growing cesium lithium borate crystals, Review of Scientific Instruments 80, 013908, 2009], American Institute of Physics.)

Sasaki et al. [81] have developed a new growth apparatus that combines propeller and crucible rotation to thoroughly stir the solution. The schematic of their apparatus is shown in Figure 9.24 [81]. CLBO crystals were grown under conditions of 30 s spin-up time at 30 rpm and 30 s spin-down time. The crucible reversed the rotation direction every 5 min. Other conditions, such as the flux composition and cooling rate, were the same as the conventional method. Crystals with dimensions of $12 \times 6 \times 5$ cm^3 were typically grown in 16 days.

Many technologically important nonlinear optical crystals, such as KTP, BBO, LBO, and BTO, are also grown by the TSSG method.

Traveling Heater/Solvent Method

The traveling heater/solvent method is a modified version of the flux method. It can also be considered as flux growth carried out using the float zone technique. In the traveling heater method, a solvent zone placed between the solid seed and the feed material is heated, and the molten zone is moved by the traveling heater. In this way, crystallization takes place at the advancing seed–solvent interface and dissolution of feed material at the solvent–feed material interface. The growth temperature used in the case of GaSb, for example, is in the range of 500°C–560°C compared to its melting point of 712°C [82]. Both Ga and In solvents have been used with the zone width 3–12 mm. The main

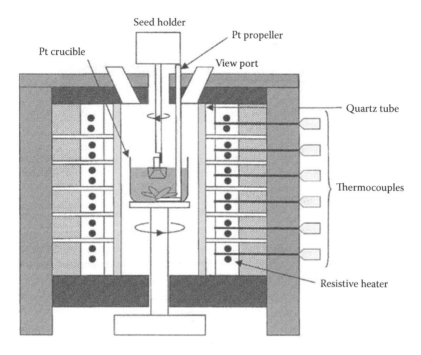

FIGURE 9.24

Schematic of the six-zone vertical furnace with a solution-stirring technique. (Reprinted from *Optical Materials*, 23, Sasaki, T., et al., Progress in the growth of a $CsLiB_6O_{10}$ crystal and its application to ultraviolet light generation, 343–351, 2003, with permission from Elsevier.)

disadvantage of this technique is its low rate of growth (a few millimeters a day), which is limited by the material transport rate of the slowest diffusing species. Several workers have used uniform rotation, accelerated rotation, or stirring by alternating magnetic fields to increase the growth rates.

With the availability of optical float zone furnaces, this method has been extended to reactive oxides. For example, Vanishri et al. [83] have grown two leg spin ladder compounds $Sr_{14}Cu_{24}O_{41}$ and $Sr_2Ca_{12}Cu_{24}O_{41}$ by the traveling solvent method using optical floating zone image furnace. Since both the parent and substituted compounds are incongruently melting, a solvent flux with excess CuO (70 mol%) was used to grow the crystals. Typically, the flux pellet is fused between the seed and feed rod and melted. The molten zone then passes through the feed rod, which dissolves the solute and crystallizes at the seed end. Figure 9.25 [83] shows the photograph of the parent compound thus grown.

General Characteristics of Flux-Grown Crystals

Since in flux growth the crystals are grown from solution, albeit at high temperature, they generally grow with their natural habits. It is worth noting

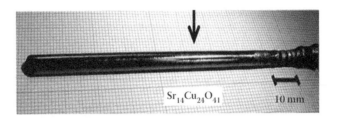

FIGURE 9.25
$Sr_{14}Cu_{24}O_{41}$ crystal grown by the traveling solvent method using optical floating zone image furnace. (Reprinted from *Journal of Crystal Growth*, 311, Vanishri, S., et al., Crystal growth and characterization of two-leg spin ladder compounds: $Sr_{14}Cu_{24}O_{41}$ and $Sr_2Ca_{12}Cu_{24}O_{41}$, 3830–3834, 2009, with permission from Elsevier.)

that the grown crystals often closely resemble their natural counterparts in habit. Habits of the flux-grown crystals and their modification have been tabulated by Wanklyn [72]. Grown crystals are seldom perfect in their form because growth is usually restricted by the crucible walls, the liquid surface, or nearby crystals. Evidence of hopper growth has also been observed in flux growth. Crystals exhibiting hopper growth generally grow at the solution interface and float while growing. Finished surfaces of flux-grown crystals often exhibit dendritic growth patterns on them. This suggests that a high degree of supersaturation results during the last stage of growth, probably during the solidification of the flux. A dendritic pattern observed on the habit face of flux-grown $SrSO_4$ is shown in Figure 9.26 [74].

It has been observed that the flux-grown crystals are generally pure. Contamination from an external source is avoided since the crucibles are usually closed or even sealed. However, the crystals cannot be claimed to be absolutely free from impurities. The main sources of impurities in flux-grown crystal are crucible material, solvent, and inherent impurities present in the solution. Solvent attack on platinum varies considerably.

Incorporation of solvent may either be as impurities in the lattice sites or in the form of gross inclusions. Contaminations at the lattice site can be minimized by choosing, whenever possible, a solvent having a common ion with the solute. Gross inclusions of the solvent are related to experimental conditions and can be minimized by proper control of various growth parameters, such as temperature stability, stirring, seeding whenever involved, and so on.

Doping with suitable impurities is relatively easy in flux growth. Doping can be simply done with the addition of dopant material in the charge itself. To increase the dopant concentration in the crystal, a larger amount of the dopant material needs to be incorporated in the charge. Since the limit of dopant concentration is solely dictated by structural considerations, the extent to which the dopant has entered the lattice will have to be determined after the growth. Doped crystals grown by slow cooling frequently show a "tailing off" of dopant concentration in the outer region. This change in

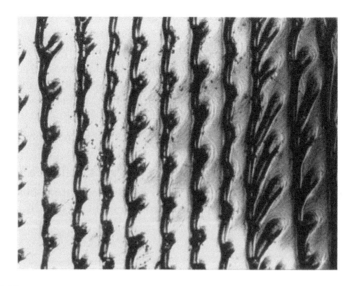

FIGURE 9.26
Dendritic pattern observed on (100) face of flux-grown SrSO$_4$ crystal. (Reprinted from *Journal of Crystal Growth*, 11, Patel, A. R., and Bhat, H.L., Growth of barite group crystals by the flux evaporation method, 166–170, 1971, with permission from Elsevier.)

concentration is rather gradual, and usually abrupt change is not found in flux-grown crystals. Even the tailing off is very small and does not pose a serious problem in the crystal's usage.

Advantages and Disadvantages

One of the primary merits of flux growth lies in the fact that it comes to the rescue of crystal growers when the material to be grown has certain undesirable properties. For example, if a material is insoluble in water, melts incongruently, decomposes on or before melting, has a high vapor pressure at the melting point, or exists as an undesirable polymorph close to the melting point, the growth has to be performed at a temperature lower than its melting point. For such crystals, molten salt solvent crystallization (i.e., the flux growth) is the most suitable method. The other advantages are as follows:

1. Although much slower compared to the melt growth, the rates of crystallization are quite rapid for a solvent–solute system.
2. Because the growth takes place well below the melting point, the flux-grown crystals are structurally more perfect than those produced by alternate methods.
3. Flux-grown crystals are usually quite pure.
4. Doping with suitable dopant materials is easily achieved.

The principal disadvantages are as follows:

1. The crystals grown from conventional flux growth are normally small. Scaling up involves considerable cost. In large vessels, crystals up to a size of ~1 cm could be obtained. However, with the invention of the TSSG methods variants of flux growth, it has now become possible to grow quite large crystals.
2. Contamination of crystals by the solvent constituent elements and sometimes that of crucible material.
3. Uniform doping is difficult because of the "tailing off" effect.
4. Even though the growth takes place at a temperature below the melting point of the material, the growth still takes place at quite high temperatures.

High-Pressure, High-Temperature Growth

The crystal growth at very high pressure and high temperature was primarily developed to produce synthetic diamond. Nature grows diamond in breathtaking sizes and quality beneath the surface of the earth. However, the growth takes place at great depths (150–160 km below the earth's surface), where the conditions of pressure and temperatures are right. The method adopted to grow diamond is essentially to mimic Mother Nature in the laboratory. Success achieved in growing diamond by this method has prompted scientists to grow cubic boron nitride, a synthetic material.

Diamond

Diamond is one of the most difficult materials to grow in single crystal form. The largest synthetic diamonds available commercially are about 1 mm^3, but larger sizes ($5 \times 4 \times 4 \text{ mm}^3$) have also been grown. Even though numerous claims have been documented on diamond growth between 1879 and 1928, the successful synthesis was first achieved in 1955 at the General Electric Company by H. T. Hall and his group [84]. The breakthrough came from using a "belt" press, which was capable of producing pressures above 100 kbar and temperatures above 2000°C. In the belt press, the upper and lower anvils apply the pressure to a cylindrical inner cell. The internal pressure thus generated is confined radially by a belt of prestressed steel bands. The anvils also serve as electrodes providing electrical current to heat the compressed cell. Figure 9.27 shows the schematic of a belt apparatus. Belt presses are still used today, but they are built on a much larger scale than those of the original design.

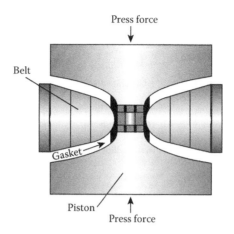

Press force

Belt

Gasket

Piston

Press force

FIGURE 9.27
Schematic of the belt press.

Although diamond can be directly synthesized from graphite at about 3000°C and 130 kbar pressure, a more practical process involves the use of a solution. The solvents used include metals like Pt, Pd, Mn, Cr, and Ta. However, the best results are obtained with Ni, Co, and Fe and their alloys. The process is essentially one of the temperature differential solution growth. To synthesize diamond by this method, graphite and the metal discs are stacked in a cylindrical cavity drilled in a pyrophyllite capsule. Pyrophyllite is a naturally occurring mineral with composition $Al_2Si_4O_{10}(OH)_2$. It is soft and easily machinable, and has excellent thermal stability. The filled cavity is closed with a pyrophyllite lid and placed in between the anvils of a suitable press, where it is subjected to high pressure and temperature. When the required pressure is reached, a heavy current (e.g., ~700 Amp at 3 volts) is passed through the cell, a pair of opposite tungsten carbide anvils carrying the current to heat the material. Heating of the capsule can be done by passing the current through the graphite sleeve surrounding the charge material, as is shown in Figure 9.28a. It can also be done directly by passing the current through the charge (Figure 9.28b). Pressure and the current are maintained for a period of time ranging from half a minute to 15–20 minutes. The current is then switched off, and the pressure slowly released. The Pyrophyllite cube is extracted. What one gets is a solid cylindrical mass, which is broken and chemically treated to remove the unconverted graphite and solvent catalyst material. Any trace of pyrophyllite is removed by treating the mass with hydrofluoric acid (HF). Figure 9.29 shows some synthetic diamonds of dodecahedral form grown by the high-pressure, high-temperature technique [85]. Table 9.4 gives some data pertaining to diamond growth.

A number of designs have been developed to apply pressure for diamond growth. These are all with opposed multianvils. The first multianvil press

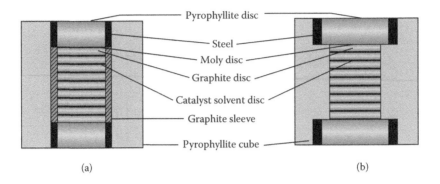

(a)   (b)

FIGURE 9.28
Filling of the pyrophyllite cavity. (a) Current through the graphite sleeve and (b) current through the charge.

FIGURE 9.29
Synthetic diamonds of dodecahedral morphology. (From Kanda, H., et al., *Journal of Crystal Growth* 60, 441–444, 1982. With permission.)

TABLE 9.4

Typical Growth Conditions

Solvent	Temperature (°C)	Pressure (kbar)	Duration of Heating (min)
Ni	1420	65–70	1/2–20
Co	1410		
Fe	1320		
–	3000	130	1/2–20

designed was a tetrahedral press, using four anvils to converge upon a tetrahedron-shaped volume [86,87]. The second type of press design is the cubic press, which was created to increase the volume to which pressure could be applied. A cubic press has six anvils, which provide pressure simultaneously onto all faces of a cube-shaped volume [88]. A cubic press is typically smaller than a belt press and can more rapidly achieve the pressure and temperature necessary to create synthetic diamond. However, cubic presses cannot be easily scaled up to larger volumes. The only way to increase the pressurized volume is to use larger anvils, but this also increases the amount of force needed on the anvils to achieve the same pressure. Presses with more number of opposed anvils have also been fabricated [88].

Cubic Boron Nitride

In contrast to crystalline forms of carbon, cubic boron nitride (c-BN), a super hard material, has no counterpart in nature and hence is purely human made. The graphite-like (hexagonal) form of boron nitride has been known for many years and is used as refractory material and lubricant. It is white in color and is an insulator, in contrast to graphite, which is black and conducting. The structural analogy between graphite and hexagonal boron nitride prompted early crystal chemists to speculate the existence of a cubic form of born nitride analogous to diamond. The speculation got realized in 1957 when Wentorf [89] first made this material at General Electric Company. Subsequently, he developed methods for the reproducible synthesis of these crystals employing high pressure and temperatures.

The procedure followed to produce c-BN is the same as that of diamond. The hexagonal boron nitride, which is the staring material, can be obtained by the reacting boron trioxide (B_2O_3) or boric acid ($B(OH)_3$) with ammonia (NH_3) or urea ($CO(NH_2)_2$) in a nitrogen atmosphere [90]. Direct conversion of hexagonal boron nitride to the cubic form is possible at pressures between 50 and 180 kbar and temperatures between 1730°C and 3230°C. However, a small amount of boron oxide when added lowers the required pressure to 40–70 Kbar and temperature to 1500°C. As in diamond synthesis, to further reduce the required pressures and temperatures, catalysts such as lithium, potassium, or magnesium; their nitrides; their fluoronitrides; and so on have also been added [91,92].

References

1. Buckley, H. E. 1951. *Crystal growth*. London, UK: Chapman and Hall.
2. Holden, A., and P. Singer. 1961. *Crystals and crystal growing*. London, UK: Heinemann.
3. Franks, F. 2000. Water: A *matrix of life*. London, UK: Royal Society of Chemistry.

4. Mullin, J. W., A. Amatavivadhana, and M. Chakraborty. 2007. Crystal habit modification studies with ammonium and potassium dihydrogen phosphate. *Journal of Applied Chemistry* 20(5): 153–158.

5. Gibbs, W. E., and W. Clayton. 1924. The production of large, clear, cubical crystals of sodium chloride. *Nature* 113: 492–493.

6. Janssen-van Rosemalen, R., W. H. van der Linden, E. Dobinga, and D. Visser. 1978. The influence of the hydrodynamic environment on the growth and the formation of liquid inclusions in large potassium dihydrogen phosphate (KDP) crystals. *Kristall und Technik* 13: 17–28.

7. Holden, A., and R. H. Thompson. 1964. *Growing crystals with a rotary crystallizer.* New York: Bell Telephone Laboratories.

8. Chary, B. R., H. L. Bhat, K. S. Sangunni, and P. S. Narayanan. 1985. Design and fabrication of a versatile solution growth system to grow large single crystals. *Journal of the Instrument Society of India* 15: 106–110.

9. Hooper, R. M., B. J. Mcardle, R. S. Narang, and J. N. Sherwood. 1980. Crystallization from solutions at low temperatures. In *Crystal growth*, edited by B. R. Pamplin. New York, NY: Pergamon Press, pp. 395–420.

10. Kruger, F., and W. Finke. 1910. Duet. Reich Pat. 228.246 K1.120 Gr. 2.

11. Walker, A. C., and G. T. Kohman. 1948. Growing crystals of ethylene diamine tartrate. *Transactions of the American Institute of Electrical Engineering* 67: 565–570.

12. Loiacono, G. M., J. J. Zola, and G. Kostecky. 1983. Growth of KH_2PO_4 crystals at constant temperature and supersaturation. *Journal of Crystal Growth* 62: 545–556.

13. Rashkovich, L. N. 1984. High—Rate growth from solutions of large crystals for nonlinear optics. *Vestnik Akademii Nauk SSSR* 9: 15–19.

14. Zaitseva, N. P., J. J. De Yoreo, M. R. Dehaven, R. L. Vital, K. E. Montgomery, M. Richardson, and L. J. Atherton. 1997. Rapid growth of large-scale (40–55 cm) KH_2PO_4 crystals. *Journal of Crystal Growth* 180: 255–262.

15. Rashkovich, L. N. 1991. *KDP-family single crystals.* Bristol, UK: Hilger.

16. Zaitseva, N. P., L. N. Rashkovich, and S. V. Bogatyreva. 1995. Stability of KH_2PO_4 and $K(H,D)_2PO_4$ solutions at fast crystal growth rates. *Journal of Crystal Growth* 148: 276–282.

17. Bordui, P. 1987. Growth of large single crystals from aqueous solution: A review. *Journal of Crystal Growth* 85: 199–205.

18. Dixit, V. K., B. V. Rodrigues, and H. L. Bhat. 2001. Experimental setup for rapid crystallization using favored chemical potential and hydrodynamic conditions. *Bulletin of Materials Science* 24: 455–459.

19. Karnal, A. K. 2006. Growth and characterization of technologically important nonlinear optical crystals cesium lithium borate and potassium di deuterium phosphate. PhD dissertation, Indian Institute of Science, Bangalore, India.

20. Sasaki, T., and A. Yokatani. 1990. Growth of large KDP crystals for laser fusion experiments. *Journal of Crystal Growth* 99: 820–826.

21. Henisch, H. K. 1970. *Crystal growth in gels.* University Park: The Pennsylvania State University Press.

22. Henisch, H. K. 1988. *Crystal growth in gels and Liesegang rings.* Cambridge, UK: Cambridge University Press.

23. Lefaucheux, F., and M. C. Robert. 1994. Crystal growth in gels. In *Hand book of crystal growth*, edited by D. T. J. Hurle. Amsterdam: North Holland, pp. 1271–1303.

24. Kalkura, S. N., and S. Natarajan. 2010. Crystallization from gels. In *Springer handbook of crystal growth*, edited by G. Dhanaraj et al. New York, NY: Springer, pp. 1607–1636.
25. Brezina, B., and M. Havrankova. 1971. Growth of KH_2PO_4 single crystals in gel. *Materials Research Bulletin* 6: 537–543.
26. Banks, E., R. Chianelli, and F. Pintchovsky. 1973. The growth of some alkaline earth orthophosphates in gelatin gels *Journal of Crystal Growth* 18: 185–190.
27. Brezina, B., M. Havrankova, and K. Dusek. 1976. The growth of $PbHPO_4$ and $Pb_4 (NO_3)_2(PO_4)_2$ $2H_2O$ in gels. *Journal of Crystal Growth* 34: 248–252.
28. Favard, F., J. P. Lechaire, M. Maillard, N. Favard, P. Andreazza, F. Lefaucheux, and M. C. Robert. 1992. 3-D-electron microscopy configuration of TMOS wet silica gels prepared by the quick-freeze, deep-etching-rotary-replication technique. *Colloid and Polymer Science* 270: 584–589.
29. Cabane, B., M. Dubois, F. Lefaucheux, and M. C. Robert. 1990. Mesh size of tmos gels in water. *Journal of Non-Crystal Solids* 119: 121–131.
30. Arora, S. K. 1981. Advances in gel growth: A review. *Progress in Crystal Growth Characterization* 4: 345–378.
31. Al Dhahir, T. A., G. Dhanaraj, and H. L. Bhat. 1992. Growth of alkali metal periodates from silica gel and their characterization. *Journal of Crystal Growth* 121: 132–140.
32. Kratochvil, P., B. Sprusil, and N. Heyrovsky. 1968. Growth of gold single crystals in gels. *Journal of Crystal Growth* 3–4: 360–362.
33. Liaw, H. M., and J. W. Faust, Jr. 1971. The dendritic growth of lead from gels. *Journal of Crystal Growth* 8: 8–12.
34. Holmes, H. N. 1926. Chapter 12. Reactions in gels. In *Colloid chemistry*, edited by J. Alexander. New York: Chemical Catalog Company, pp. 118–126.
35. Armington, A. F., and J. J. O'Connor. 1968. Gel growth of cuprous halide crystals. *Journal of Crystal Growth* 3–4: 367–371.
36. Glocker, D. A., and I. F. Soest. 1969. Growth of single crystals of monobasic ammonium phosphate in gel. *Journal of Chemical Physics* 51: 3143–3143.
37. Frank, F. C. 1950. Radially symmetric phase growth controlled by diffusion. *Proceedings of the Royal Society of London A* 201: 586–599.
38. Bhat, H. L. 1981. Filamentary and dendritic growth of lead chloride crystals in silica gel. *Journal of Materials Science* 16: 1707–1710.
39. Bhat, H. L. 1973. Studies on growth and defect properties of barite group crystals. PhD dissertation, Sardar Patel University, VallabhVidynagar, India.
40. Ives, M. B., and J. Plewes. 1965. Inhibited dissolution of {100} surfaces of single crystals of lithium fluoride. *Journal of Chemical Physics* 42: 293–295.
41. Patel, A. R., and A. V. Rao. 1979. Gel growth and perfection of orthorhombic potassium perchlorate single crystals. *Crystal Growth* 47: 213–218.
42. Patel, A. R., and A. V. Rao. 1982. Crystal growth in gel media. *Bulletin of Materials Science* 4: 527–548.
43. Byrappa, K. 1994. Hydrothermal growth crystals. In *Handbook of crystal growth*, edited by D. T. J. Hurle. Amsterdam: North Holland, pp. 465–562.
44. Byrappa, K. 2010. Hydrothermal growth of polyscale crystals. In *Springer handbook of crystal growth*, edited by G. Dhanaraj et al. New York, NY: Springer, pp. 599–653.
45. Laudise, R. A., and J. W. Nielsen. 1961. Hydrothermal crystal growth. In *Solid state physics*, Vol. 12, edited by F. Seitz and D. Turnbull. New York: Academic Press, pp. 149–222.

46. Morey, G. W., and P. Niggli. 1913. The hydrothermal formation of silicates, a review. *Journal of the American Chemical Society* 35: 1086–1130.
47. Roy, R., and O. F. Tuttle. 1956. Investigations under hydrothermal conditions. *Physics and Chemistry of Earth* 1: 138–180.
48. Walker, A. C., and E. Buehler. 1950. Growing large quartz crystals. *Industrial and Engineering Chemistry* 42: 1369–1375.
49. Bridgman, P. W. 1914. The technique of high pressure experimenting. *Proceedings of the Academy of Arts and Sciences* 49: 627–643.
50. Ballman, A. A., and R. A. Laudise. 1963. Hydrothermal growth. In *The art and science of growing crystals*, edited by J. J. Gilman. New York, NY: John Wiley, pp. 231–251.
51. Kennedy, G. C. 1950. Pressure-volume-temperature relations in water at elevated temperatures and pressures. *American Journal of Science* 248: 540–564.
52. Kolb, E. D., P. L. Key, and R. A. Laudise. 1983. Pressure-volume-temperature behavior in the system H_2O-NaOH-SiO_2 and its relationship to the hydrothermal growth of quartz. In *Proceedings of the 37th Annual Symposium on Frequency Control*, pp. 153–156.
53. Spezia, G. 1905. La pressione è chimicamente inattiva nel la solubilità e ricostituzione del quarzo [Pressure is chemically inactive in the solubility and reconstitution of quartz]. *Atti della Reale Accademia delle scienze di Torino [Proceedings of the Royal Academy of Sciences in Turin]* 40: 254–262.
54. Laudise, R. A., and A. A. Ballman. 1961. The solubility of quartz under hydrothermal conditions. *Journal of Physical Chemistry* 65: 1396–1400.
55. Hosaka, M., and S. Taki. 1981. Hydrothermal growth of quartz crystals in NaCl solution. *Journal of Crystal Growth* 52: 837–842.
56. Lafon, F., and G. Demazean. 1994. Pressure effects on the solubility and crystal growth of α-quartz. *Journal de Physique* 4(C2): 177–182.
57. Nassau, K. 1976. Synthetic emerald: The confusing history and the current technologies. *Journal of Crystal Growth* 35: 211–222.
58. Laudise, R. A., and E. D. Kolb. 1962. Hydrothermal crystallization of yttrium-iron garnet on a seed. *Journal of the American Ceramics Society* 45: 51–53.
59. Distler, G. I., S. A. Kobzareva, A. N. Lobachev, O. K. Melnikov, and N. S. Triodina. 1978. On the growth mechanism of sodalite single crystals grown by the hydrothermal method on single crystal seeds coated with informative informative interfacial layers. *Kristal und Technik* 13: 1025–1034.
60. Ikornikova, N. Y. 1975. *Hydrothermal synthesis of crystals in chloride system.* Moscow: Nauka, pp. 1–122.
61. Kolb, E. D., and R. A. Laudise. 1978. Hydrothermal synthesis of aluminum orthophosphate. *Journal of Crystal Growth* 43: 313–319.
62. Byrappa, K., and K. V. K. Shekar. 1992. Hydrothermal synthesis and characterization of piezoelectric lithium tetraborate, li2b4o7, crystals. *Journal of Material Chemistry* 2: 13–18.
63. Laudise, R. A., W. A. Sunder, R. F. Belt, and G. Gashurov. 1990. Solubility and P-V-T relations and the growth of potassium titanyl phosphate. *Journal of Crystal Growth* 102: 427–433.
64. Remeika, J. P. 1954. A method for growing barium titanate single crystals. *Journal of the American Chemical Society* 76: 940–941.
65. Elwell, D., and H. J. Scheel. 1975. *Crystal growth from high temperature solution.* London, UK: Academic Press.

66. Laudise, R. A. 1963. Molten salt solvents. In *The art and science of growing crystals*, edited by J. J. Gilman. New York, NY: John Wiley, pp. 252–273.

67. White, E. A. D. 1965. Growth of single crystals from the fluxed melt. In *Techniques of inorganic chemistry*, Vol. 4, edited by H. B. Jonassen and A. Weissberger. New York: John Wiley, pp. 31–63.

68. Tolksdorf, W. 1994. Flux growth. In *Handbook of crystal growth*, Vol. 2, edited by D. T. J. Hurle. Amsterdam, The Netherlands: North Holland, pp. 263–611.

69. Carvajal, J. J., M. C. Pujol, and F. Diaz. 2010. High temperature solution growth: Application to laser and nonlinear optical crystals. In *Springer handbook of crystal growth*, edited by G. Dhanaraj et al. New York, NY: Springer, pp. 725–757.

70. Levin, E. M., C. R. Robbins, and H. F. McMurdie. 1964. *Phase diagrams for ceramists*. Westerville, OH: American Ceramics Society. (see also, supplements in 1969, 1975, 1981, 1983, 1984).

71. Nielsen, J. F., and R. R. Monchamp. 1970. The use of phase diagrams in crystal growth. In *Phase diagrams*, Vol. 3, edited by A. M. Alper. New York, NY: Academic Press.

72. Wanklyn, B. M. 1974. Practical aspects of flux growth by spontaneous nucleation. In *Crystal growth*, Vol. 1, edited by B. R. Pamplin. Oxford, UK: Pergamon Press, pp. 217–288.

73. Tolksdorf, W. 1977. Growth of magnetic garnet single crystals from high temperature solution. *Journal of Crystal Growth* 42: 275–283.

74. Patel, A. R., and H. L. Bhat. 1971. Growth of barite group crystals by the flux evaporation method. *Journal of Crystal Growth* 11: 166–170.

75. Webster, F. W., and E. A. D. White. 1969. Solution growth of magnesium oxide crystals. *Journal of Crystal Growth* 5: 167–170.

76. Hintzmann, W., and G. Muller-Vogt. 1969. Crystal growth and lattice parameters of rare-earth doped yttrium phosphate, arsenate and vanadate prepared by the oscillating temperature flux technique. *Journal of Crystal Growth* 5: 274–278.

77. Scheel, H. J., and D. Elwell. 1972. Stable growth rates and temperature programming in flux growth. *Journal of Crystal Growth* 12: 153–161.

78. Linz, A., V. Belruss, and C. S. Naiman. 1965. Electrochemical Society Spring Meeting, San Fransisco, Extended Abstracts 2: 87.

79. Sasaki, T., Y. Mori, I. Kuroda, S. Nakajima, K. Yamaguchi, S. Watanabe, and S. Nakai. 1995. Caesium lithium borate: A new nonlinear optical crystal. *Acta Crystallographica Section* C 51: 2222–2224.

80. Babu Reddy, J. N., E. Suja, H. L. Bhat, and A. K. Karnal. 2009. Development of a versatile high temperature top seeded solution growth unit for growing cesium lithium borate crystals. *Review of Scientific Instruments* 80: 013908.

81. Sasaki, T., Y. Mori, and M. Yoshimura. 2003. Progress in the growth of a $CsLiB_6O_{10}$ crystal and its application to ultraviolet light generation. *Optical Materials* 23: 343–351.

82. Danilewsky, A. N., S. Lauer, J. Meinhardt, K. W. Benz, B. Kaufmann, R. Hofmann, and A. Dornen. 1966. Growth and characterization of GaSb bulk crystals with low acceptor concentration. *Journal of Electronic Materials* 25: 1082–1087.

83. Vanishri, S., C. Marin, H. L. Bhat, B. Salce, D. Braithwaite, and L. P. Regnault. 2009. Crystal growth and characterization of two-leg spin ladder compounds: $Sr_{14}Cu_{24}O_{41}$ and $Sr_2Ca_{12}Cu_{24}O_{41}$. *Journal of Crystal Growth* 311: 3830–3834.

84. Bundy, F. P., H. T. Hall, H. M. Strong, and R. H. Wentorf. 1955. Man-made diamonds. *Nature* 176: 51–54.

85. Kanda, H., N. Setaka, T. Ohsawa, and O. Fukunaga. 1982. Growth condition for the dodecahedral form of synthetic diamond. *Journal of Crystal Growth* 60: 441–444.
86. Hall, H. T. 1958. Ultrahigh pressure research. *Science* 128: 445–449.
87. Hall, H. T. 1960. Ultrahigh pressure, high temperature apparatus: The belt. *Review of Scientific Instruments* 31: 125–131.
88. Ito, E. 2007. Multi-anvil cells and high-pressure experimental methods. In *Treatise of geophysics*, edited by G. Schubert. Amsterdam, The Netherlands: Elsevier, pp. 197–230.
89. Wentorf, R. H. 1957. Cubic form of boron nitride. *Journal of Chemical Physics* 26: 956–957.
90. Rudolph, S. 2001. Boron nitride (BN) in materials review: Alumina to zirconia. *American Ceramic Society Bulletin* 80(8): 68–71.
91. Vel, L., G. Demazeau, and J. Etourneau. 1991. Cubic boron nitride: Synthesis, physic-chemical properties and applications. *Materials Science and Engineering B* 10: 149–164.
92. Fukunaga, O. 2002. Science and technology in the recent development of boron nitride materials. *Journal of Physics: Condensed Matter* 14(44): 10979–10982.

10

Crystal Growth from the Vapor Phase

Introduction

Until the 1960s, vapor growth techniques were mainly used to obtain bulk crystals. With the ever-increasing importance of thin films in technological applications, a range of vapor phase techniques, some of which are extremely sophisticated, have been developed to prepare epitaxial thin films and are being used extensively. With this changed situation, it looked as though bulk growth of crystals from the vapor phase was no longer relevant. However, the technique has got a big boost in recent years by the great demand for SiC crystals, which are industrially produced by the vapor transport technique. In this chapter, we deal with those techniques that are used to prepare bulk single crystals. In contrast to the epitaxial growth methods, which we discuss in Chapter 11, growth of bulk crystals from the vapor phase is carried out at relatively high pressures. Consequently, the mean free path of the vapor molecules is much smaller here and hence the vapor can be considered as a fluid.

Several general classes of techniques are available for the growth of bulk crystals from the vapor phase. They have been reviewed in the past by experts in the field [1–3]. The early works have been summarized in *The Art and Science of Growing Crystals,* the book edited by Gilman [4]. Growth of single crystals from the vapor phase with special reference to semiconductors has been reviewed recently by Dhanasekaran [5]. Vapor growth can either be monocomponent or polycomponent. Obviously, crystals grown from monocomponent techniques have less contamination, but the growth will occur at higher temperature. On the other hand, in polycomponent techniques, the growth occurs at lower temperature, but the techniques have the disadvantage of contamination and diffusion-related problems due to the presence of additional component(s).

Since the growth rate possible in vapor phase techniques is in the range of millimeters per day, which is at least an order of magnitude lower than those achievable in melt-based growth techniques, these are used only when the melt-based techniques are not applicable. For the same reason, the crystals grown from the vapor phase contain lower concentration of defects.

Classification

The following chart gives a broad classification of the available vapor phase growth techniques. This is not a unique classification, but it is useful in presenting the subject in certain order. Thin film techniques have not been classified further in this chart. As can be inferred from the chart, material transport in the vapor form is the essence of all bulk growth techniques.

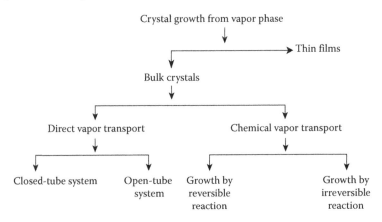

Vapor Transport Mechanisms

In vapor phase techniques carried out in isothermal environment, the material transport is mainly governed by Fick's law of diffusion, neglecting the effect of gravity. However, where the crystals grow as a result of temperature gradient, several transport mechanisms may be operative depending on the experimental situation. To avoid secondary nucleation, the growth parameters are generally set such that the gas layers close to the growing crystal are only slightly supersaturated, thereby maintaining a quasi-equilibrium condition in the vicinity of the growth front. Obviously, large temperature differences and fluctuations are to be avoided. We consider next the operative transport mechanisms.

Convective Flow

Convective flow is produced in gases by the combined effect of spatial density differences and gravity. This is, however, dominant in systems that are of large dimensions and fast flow. If the dimensions are small (say, a few centimeters), convection may be neglected because the variation in density of gases is relatively small. In the limit of small spatial variation in density, the flow can be regarded as laminar.

Laminar Flow

This occurs when the total pressure difference between the feed materials and the growing crystal is very small, which brings in orderliness in mass transfer. At low velocities, the gases tend to flow without lateral mixing, which means there is no crosscurrent perpendicular to the direction of flow. The correlation between the total pressure difference and the laminar flux is given by the Hagen–Poiseuille law, which is written as

$$J = \pi r^4 \Delta P/(8\eta L) \tag{10.1}$$

where J is the volumetric flow rate, r and L are the tube radius and length, ΔP is the pressure drop (including that due to any gravity head) in the direction of flow, and η is the fluid viscosity. Substitution of suitable values corresponding to the experimental situation into Equation 10.1 shows that very small pressure differences can give rise to mass transfer [2]. In case of larger flow due to larger temperature gradient, the actual fluid flow will no longer be laminar but turbulent.

The average drift velocity of the gas flow is given by $v = J/d$, where d is the density of the vapor. Then, the flux due to laminar flow can be written as

$$J^{lam} = v.d^{lam} = v.P^{lam}/RT \tag{10.2}$$

where R is the gas constant and T is the temperature.

The rate of the laminar flow will also be limited by the presence of inert gases or volatile impurities through diffusion, which we discuss in the next section.

Diffusion

Diffusion occurs if the system contains inert gases or gaseous impurities besides the main component. In this case, the minority component has to diffuse through the concentration gradient of the main component. This results in a concentration gradient of the minor component also, its concentration being lower at the growing front. Due to these concentration gradients, a diffusion flux is generated and the total flux will be the sum of the laminar and diffusive fluxes. Consequently, the growth becomes diffusion controlled.

The diffusive flux can be written as

$$J^{diff} = (-D/RT)\, dP/dx \tag{10.3}$$

where D is the diffusion coefficient.

Thermodiffusion (Soret Effect)

Thermodiffusion is a process through which the separation of the components of a mixture toward hot/cold regions would occur, again because of the imposed temperature gradient. According to Factor and Garrett [1], this may be neglected, except in systems containing a large quantity of H_2 gas.

Total Flux

Let us consider the case of a gaseous species α with a vapor pressure P_α diffusing through an inert gas β, which is at a pressure P_β. Further, for simplicity, we treat this as a one-dimensional problem. In a system where transport is composed of laminar flow and diffusive flow, the total flux ($J^{tot} = J^{lam} + J^{diff}$) of the species α is

$$J_\alpha = v(P_\alpha)^{lam}/RT + (-D_{\alpha\beta}/RT)\, dP_\alpha/dx \qquad (10.4)$$

The flux of the inert gas on the other hand, is

$$J_\beta = v(P_\beta)^{lam}/RT + (-D_{\beta\alpha}/RT)\, dP_\beta/dx = 0 \qquad (10.5)$$

The total pressure in the system is $P = P_\alpha + P_\beta$, and since the pressure gradient is small, we can assume $dP/dx \approx 0$ so that

$$v \approx J_\alpha RT/P \qquad (10.6)$$

Substituting this for v in Equation 10.4 and rearranging, we obtain

$$J_\alpha = -(D_{\alpha\beta}/RT)[P/(P - P_\alpha)]dP_\alpha/dx \qquad (10.7)$$

Stefan [6] first proposed this relationship while studying the evaporation of liquids in pipes, which was experimentally verified by him. This relationship is central to the understanding of physical vapor transport and is now called Stefan flow. It can be used for calculating the flux and hence the concentration of the species as a function of transport distance.

It must be noted that the aforementioned problem is an oversimplified one. Numerical solution to the corresponding problem in two dimensions has shown the importance of viscous interaction in the flow. Solution confirms the presence of core flow of the second component, which was considered stagnant in one-dimensional case [7]. When wider ampoules are used, convective flow cannot be neglected.

Mass transport mechanism in systems where a chemical reaction is also involved is much more complicated. Here, transport must take place in two opposite directions: a volatile intermediate species from source to crystal and the transporting agent from crystal to source. These have been discussed in greater details in some of the review articles cited earlier in this chapter.

Nucleation Control

A major problem encountered in the growth of bulk crystals from the vapor phase is nucleation control, which is particularly relevant when growth occurs in the absence of a seed. Since the condensation occurs on the inner wall of the glass or silica ampoule, it is normally flame polished to close any micro pores or striations on the surface that would otherwise act as nucleation centers. Temperature oscillation method analogous to the one described in the flux growth has also been applied in vapor growth wherein the periodic growth and resublimation occur, thereby removing all but a few large crystals to grow further. Further, nucleation control can be achieved by localizing the nucleation region within the ampoule by attaching a cold finger. For example, large crystals of ZnS have been grown by cooling the growth end of the ampoule by passing cold air through an attached tube [8]. Localizing nucleation can also be done on the side wall of the growth ampoules by providing a cold spot there [9].

Kaldis and his group have provided three guidelines for effective nucleation control [10–13]. Following these guidelines, many successful experiments have been performed to grow bulk crystals, which have been described in the reviews cited earlier. The guidelines are as follows:

1. The imposed supersaturation should be subcritical so that nucleation would be initiated only on a few active sites on the ampoule wall. Careful design of the tapered ampoule tip can further aid to achieve this objective, as is done in the Bridgman technique.

2. The ampoule traverse rate through the temperature gradient should be matched with the linear growth rate of the crystal to ensure a constant supersaturation at the growth front throughout the growth run. This will avoid spurious nucleation at later stages of growth.

3. Ampoule should be cylindrical and its diameter should be large enough to promote convective mixing.

It is worth noting that nucleation control by seeding is not widely practiced in vapor growth of bulk crystals, particularly at an exploratory level. This is probably because of the difficulties involved in seeding. However, seeded growth obviously offers the advantage of control over the orientation of the growing crystal. For example, silicon carbide crystals are usually grown on a seed crystal plate of known orientation.

Furnaces

Many vapor growth experiments in closed systems are performed in vertical gradient furnaces with cylindrical cross-section in what is known as an inverted Bridgman configuration. These have the advantages of symmetrical radial temperature distribution as well as ease of ampoule traverse as compared to horizontal furnaces. Disadvantage of vertical configuration is the convective flow in the furnace tube that can smear the thermal profile. When the ampoules are kept stationary, horizontal gradient furnaces are quite satisfactory. Expensive, transparent silica furnaces with evaporated semitransparent gold film as radiation shield have also been used [14].

Experimental Techniques

Direct Vapor Transport

The solid–vapor, solid–liquid, and liquid–vapor equilibrium curves of certain substances intersect at the triple point T_t, as shown in Figure 10.1, where it coexists in all the three phases in equilibrium. The solid–vapor equilibrium curve in this P–T diagram is called the sublimation curve. Techniques we discuss in this section refer to the region marked by the horizontal dotted line, whereas those we discussed under growth from melt refer to the region marked by the horizontal dashed line.

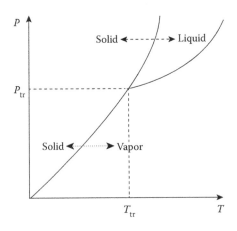

FIGURE 10.1
P–T diagram showing sublimation curve.

In direct vapor transport, the growth is carried out either in a closed-tube or in an open-tube system. In a closed-tube system, the material on sublimation is transported directly, the driving force being the temperature gradient. In an open-tube system, a carrier gas is additionally used to transport the species from the source region to the growth region.

Closed-Tube System

Conceptually, the sublimation technique is the simplest to visualize. Here, the source material is kept at one end of the sealed tube and heated to a temperature so that it sublimes. The vapor is transported to the other end of the tube, which is at a lower temperature where it condenses to form crystallites. The transport is initiated by a difference in the partial pressures of the species to be transported. The movement is from higher to lower partial pressure, which for a sublimation reaction is from higher temperature region to lower temperature region. This is schematically shown in Figure 10.2 for horizontal configuration.

This method was first used by Pizzarello [15] for the growth of lead sulfide crystals. Since then, several investigators have used this method to grow crystals in either of the configurations.

If the source material dissociates on heating, then the transport mechanism becomes more complex. In crystals grown by dissociative sublimation, the stoichiometry is an issue that requires further consideration. The extra amount to be added to control the vapor stoichiometry is usually so small that accurate control becomes impossible. Several innovative methods have been employed to overcome this problem, chief among them being to keep a separate reservoir for the more volatile component whose temperature could be independently controlled, as has been done for selenium in the growth of lead selenide [16] and zinc selenide [17].

Figure 10.3 shows the schematic of the ampoule shape used to grow crystals in the vertical furnace configuration. Here, the filled ampoule is slowly pulled through a temperature gradient.

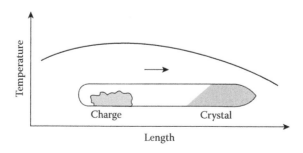

FIGURE 10.2
Sublimation technique: horizontal configuration.

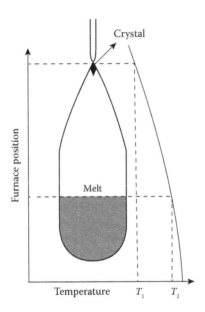

FIGURE 10.3
Direct vapor transport in inverted Bridgman configuration.

Ideally, the pulling rate should match the linear growth rate of the growing crystal. Some classic examples of crystals grown by direct vapor transport are CdS, ZnSe, Cd_4GeS_6, HgI_2, CdTe, CdZnTe, and so on.

Growth of Silicon Carbide

The applicability of the direct vapor transport technique to grow large crystals is best illustrated in the case of growth of silicon carbide (SiC). In fact, SiC is one of the oldest semiconductor materials known and has received special attention in recent times. This is primarily because of its suitability in high-power, high-frequency, and high-temperature electronic and optoelectronic applications. The material is also highly radiation resistant and hence can be used in space application. Consequently, there are renewed efforts in growing this crystal. Unlike most semiconductor crystals, SiC cannot be grown from melt with easily achievable growth conditions. Calculation has shown that it would melt at above 3200°C at a pressure above 10,000 atm [18]. However, because SiC readily sublimes (sublimation temperature ~2700°C), it can be grown by vapor phase techniques. In fact, vapor phase growth has actually become the primary method for obtaining large SiC crystals. SiC exists in about 250 polymorphic forms. Alpha silicon carbide (α-SiC) is the most commonly encountered polymorph. It has a hexagonal crystal structure and is designated as 6H. Other prominent polymorphs are 4H (hexagonal) and 3C (cubic zinc blend structure).

Although commercial production of polycrystalline SiC was established as early as 1892 by employing the Acheson method [19] and small single crystals could be extracted from the product, Lely in 1955 [20] was the first to demonstrate the single crystal growth of SiC from the vapor phase and the method is named after him. In the Lely method, lumps of SiC are packed between two concentric graphite tubes, after which the inner tube is carefully removed, leaving a thick layer of porous SiC on the inside wall of the outer graphite tube. The open ends of the tube with the charge are closed with graphite or SiC lids and loaded vertically into a furnace. The furnace is then heated to a temperature of about 2500°C in an argon (inert) atmosphere of 1 bar.

SiC closer to the graphite tube sublimes and the vapor slowly flows toward the inner exposed surface where crystals start to nucleate by condensation and grow as thin hexagonal platelets. This happens because the charge is porous and its inner exposed wall is at a slightly lower temperature than the graphite tube. If the heating is prolonged, the platelets grow larger. However, in this process, there is no control over nucleation and the grown crystals would not be very large. Figure 10.4a schematically represents the Lely process.

In 1978, Tairov and Tsvetkov [21] developed the seeded sublimation growth technique to grow SiC, which is commonly referred to as the modified Lely method. In this method, the growth again occurs in an argon environment of 1 bar, but on a seed plate. The growth takes place in a temperature range of 1800°C–2600°C, and the vapor transport is facilitated by the difference in temperature between the source and seed, the seed being at a slightly lower temperature. In a commonly used configuration, which is due to Barrett et al. [22], the source material is placed at the bottom of the cylindrical graphite crucible and the seed plate on the top. Figure 10.4b shows the schematic of

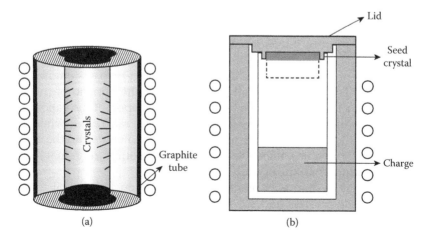

(a) (b)

FIGURE 10.4
Growth of SiC by sublimation: (a) Lely's method and (b) modified Lely's method.

FIGURE 10.5
SiC boule grown at C-MET, Hyderabad, India. (From Mahajan, S., et al., *Growth of 2" silicon carbide (SiC) single crystal*, A paper presented at the 25th Annual Meetings of the Materials Research Society, India, 2014. With permission.)

this configuration. Using a seed of 33 mm diameter, they could grow 18 mm thick crystal. Because this arrangement has high yield, it has been adopted by industry to grow large boules. Currently, crystals are grown at the rate of about 1 mm/h. Readers may refer to a chapter written by Dhanaraj et al. [23] in the *Springer Handbook on Crystal Growth* to know more about the growth and characterization of SiC. Figure 10.5 shows a SiC boule grown in C-MET, Hyderabad, India [24].

Open-Tube System

The schematic of the vapor phase growth in an open-tube system is shown in Figure 10.6. The source material is vaporized at the hot end of the tube, and an inert gas flows through the tube causing transport of the vapor molecules from the higher temperature to the lower temperature side where they deposit. If volatile components are used as starting materials and the desired compound is formed at a higher temperature, the source zone may be at a lower temperature than the crystallization zone.

The open-tube system was pioneered by Piper and Polich [25] to grow cadmium sulfide crystals and subsequently has been used to grow many other crystals such as ZnS, CdSe, and ZnSe. Figure 10.7 shows a CdS crystal grown in a specially designed open-tube system [25].

Gas flow–assisted vapor transport method has also been employed to grow C_{60} crystals. For example, using high-purity helium as a carrier gas,

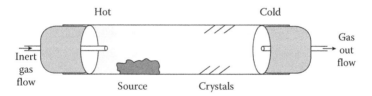

FIGURE 10.6
Open-tube system.

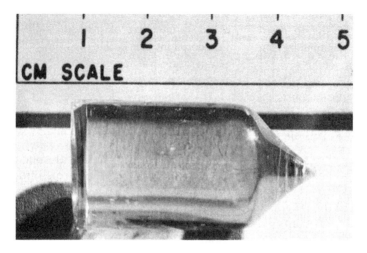

FIGURE 10.7
CdS crystal grown by open-tube system. (Reprinted with permission from [Piper, W. W., and Polich, S. J., Vapor phase growth of single crystals of II -VI compounds, *Journal of Applied Physics*, 32, 1278–1279, 1961], American Institute of Physics.)

well-shaped crystals as large as 2 mm along the edges have been grown in a typical open-tube system [26]. Here, the starting material, pure C_{60} powder, was placed in a gold boat at one end of the quartz cylindrical tube and heated to a temperature in the range of 590°C–620°C. The cooler part of the tube where the crystallization occurred was maintained at a temperature slightly below 520°C. The gas flow rate during the growth was below 5 cc/min, and the growth lasted for 3–5 days.

Open-tube systems have significant advantages. They are easily demountable; hence, it is easier to remove the grown crystals. Also the charge can be introduced into the system just before the commencement of the growth run. The main advantage of the open-tube system, however, is the controllability it provides in terms of doping at different stages and on transport rate. The wastage of raw materials that escape the growth chamber is the main disadvantage of this method; hence, the technique is not favored by commercial growers.

Chemical Vapor Transport

The feature common to this group of methods is that they all involve a chemical reaction. The chemical reactions may be reversible or irreversible. They may be simple or complex. But owing to chemical reaction, the concentration of the substances transported in the vapor phase increases sharply; hence, it is possible to crystallize even those materials that have negligible vapor pressure at crystallization temperatures. In fact, the introduction of vapor transport aided by chemical reaction is considered as one of the most important contributions to the field of crystal growth, which has tremendously enhanced the scope of vapor phase growth. Consequently, literature pertaining to chemical vapor transport is prolific. We discuss here only a few classic examples.

Growth by Reversible Reaction

In this case, the crystallizing substance chemically reacts with solid or liquid form of another substance in the source zone and transforms into gaseous compound. This is then transported to a zone, which is at a different temperature where the reverse reaction takes place. Transporting agents are taken only in minute quantity as they are not consumed but are only recycled and hence available for further reaction.

The reversible transport reaction process was pioneered by van Arekel and de Boer [27] to prepare high-purity refractory metal Zr (melting point: 1852°C) by what is sometimes called the hot wire process. A similar process was used to prepare Ti (melting point: 1668°C). Although the process was originally meant for obtaining pure material, it can be used to produce very good single crystals as well.

To obtain single crystals through this process, single crystal wire (say, prepared by the strain-annealing technique) of the desired material is taken as a filament in a reactor whose inner walls are lined with source material (crude Zr for Zr crystal). The reactor is to be evacuated initially to prevent the oxidation of Zr. The reactor has an additional reservoir in which I_2 or ZrI_4 is placed as shown in Figure 10.8 schematically. The partial pressure of I_2 or ZrI_4 is regulated by controlling the temperature of the reservoir. The reactor is configured in such a way that when the filament is heated to ~1200°C, the Zr lining attains a temperature of ~450°C by radiation heating. This leads to a forward reaction taking place at the lining and a reverse reaction at the filament as per the following reaction, with Zr deposited on the wire.

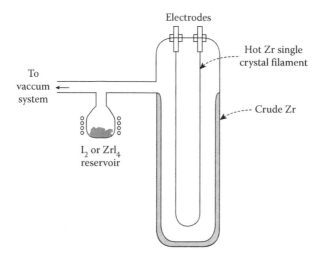

FIGURE 10.8
Schematic of the apparatus used in the hot wire process.

$$Zr(s) + 2I_2(g) \underset{1200°C}{\overset{450°C}{\rightleftharpoons}} ZrI_4(g)$$

Moliere and Wagner [28] grew single crystals of zirconium and tungsten using the hot wire process. They used tungsten wire with short barbs in both the cases. The tips of the barbs acted as nucleation centers. The barbs were recrystallized prior to growth. For the growth of tungsten crystals, the wire was maintained at 1800°C–2000°C, whereas the source material WBr_5 was at 300°C. Platy crystals of 3 mm in lateral dimension were formed at the tips of the barbs in about 8–10 h. During this period thickness of the wire also increased from 1 mm to about 4 mm.

Other examples of reversible reactions are

$$2CdS(s) + 2I_2(s) \underset{400°C}{\overset{1000°C}{\rightleftharpoons}} 2CdI_2(g) + S_2(g)$$

$$Al_2O_3(s) + 2H_2(g) \underset{1500°C}{\overset{1800°C}{\rightleftharpoons}} Al_2O(g) + 2H_2O(g)$$

$$Si I_4(g) + Si(s) \underset{900°C}{\overset{1100°C}{\rightleftharpoons}} 2SiI_2(g)$$

Growth by Irreversible Reaction

Growth by irreversible reactions is mostly carried out in an open-tube system. Even though the possibility of reverse reaction cannot be ruled out here, the reactants are not recycled in this method of growth but usually sent out of the system.

Growth of palladium whiskers by decomposition reaction is the simplest to understand [29]. The decomposition reaction is as follows:

$$Pd\ Cl_2 \rightarrow Pd + Cl_2$$

In the growth experiment, solid $PdCl_2$ taken in a vitreous silica tube is placed inside a gradient furnace and heated to a temperature between 860°C and 1000°C. The reaction is carried out by heating $PdCl_2$ rapidly in slowly flowing argon gas. Since $PdCl_2$ decomposes at 500°C, the vapor pressure is appreciable at this temperature. Growth of whiskers is initiated with the slow development of protuberances in the crust-like initial deposit of palladium from which whiskers suddenly appear and grow to a size of 2 cm. Also platelets with hexagonal and triangular morphologies were formed on the top and lateral surfaces of the whiskers.

Other typical reactions are given as follows:

$$SiH_4(g)(silane) \rightarrow Si(s) + 2H_2(g)$$

(Here, silane is pyrolyzed according to the above reaction when it is flown onto a heated Si substrate at temperatures from 950°C to 1250°C.)

$$Cd(g) + H_2S(g) \rightarrow CdS(s) + H_2(g)$$

(In this system, cadmium vapor is carried by a hydrogen stream to a reaction zone, where it reacts with a H_2S stream to form CdS. The schematic of the reactor is shown in Figure 10.9. Crystals grow as thin platelets or ribbons.)

It has also been possible to grow refractory oxides in an open tube employing the irreversible reaction method. For example, Schaffer [30] grew sapphire crystals in an open tube using the following reaction.

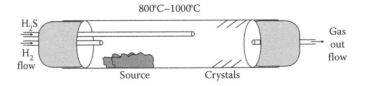

FIGURE 10.9
Growth of CdS by irreversible reaction process.

$$2AlCl_3(g) + 3H_2(g) + 3CO_2(g) \rightarrow Al_2O_3(s) + 3CO_2(g) + 6HCl(g)$$

Oriented crystals weighing up to 5 g were grown on a seed substrate in the temperature region 1550°C–1800°C. He found that the growth rate for constant gas compositions and flow rates increased with an increase in temperature.

Concluding Remarks

In this chapter, only the essentials of vapor phase growth have been discussed without attempting to cover the latest developments. As mentioned earlier, while the early attempts have been to grow the bulk crystals, the search for axial screw dislocations in whiskers gave an impetus to grow a variety of whiskers by the vapor phase techniques. The status of vapor phase technique is further enhanced due to its adaptation by the industry to grow bulk SiC crystals. However, extensive application of vapor phase techniques is in the growth of epitaxial films of semiconductors, which we discuss in Chapter 11.

References

1. Factor, M. M., and T. Garrett. 1974. *Growth of crystals from the vapor*. London, UK: Chapman & Hall.
2. Kaldis, E. 1974. Principles of vapor growth of single crystals. In *Crystal growth*, Vol. 1, edited by C. H. L. Goodman. London, UK: Plenum Press, pp. 49–191.
3. Kaldis, E., and M. Piechotka. 1994. Bulk crystal growth by physical vapor transport. In *Handbook of crystal growth*, Vol. 2a, edited by D. T. J. Hurle. Amsterdam: North Holland, pp. 613–658.
4. Gilman, J. J., ed. 1963. *The art and science of growing crystals*. New York: John Wiley.
5. Dhanasekaran, R. 2010. Growth of semiconductor crystals from vapor phase. In *Springer handbook of crystal growth*, edited by G. Dhanaraj et al. New York, NY: Springer, pp. 897–935.
6. Stefan, J. 1890. Uber die Verdampfung und die Auflosung als Vorgange der Diffusion. *Annalen der Physik* 277: 727–747.
7. Greenwell, D. W., B. L. Markham, and F. Rosenberger. 1981. Numerical modeling of diffusive physical vapor transport in cylindrical ampoules. *Journal of Crystal Growth* 51: 413–425.
8. Nitsche, R. 1967. Kristall zucht aus der Gasphase durch chemische Transport reaktionen. *Fortschritte der Mineralogie* 44: 231.
9. Honigmann, B. 1954. Züchtung größerer Einkristalle von Hexamethylentetramin aus der Dampfphase. *Zeitschrift für Elektrochemie*. 58: 322–327.
10. Kaldis, E. 1965. Keimbildung und Wachstum von grossen Einkristallen durch Transport-Reaktionen. II. ZnSe. *Helvetica Physica Acta* 38: 357–357.

11. Kaldis, E., and R. Widmer. 1965. Nucleation and growth of single crystals by chemical transport—I cadmium-germanium sulphide. *Journal of Physics and Chemistry of Solids* 26: 1697–1700.
12. Kaldis, E. 1965. Nucleation and growth of single crystals by chemical transport—II zinc selenide. *Journal of Physics and Chemistry of Solids* 26: 1701–1705.
13. Kaldis, E. 1969. Crystal growth and growth rates of CdS by sublimation and chemical transport. *Journal of Crystal Growth* 5: 376–390.
14. Reed, T. B. 1969. Solid state research report. *MIT Report* 1: 21.
15. Pizzarello, F. 1954. Vapor phase crystal growth of lead sulfide crystals. *Journal of Applied Physics* 25: 804–805.
16. Prior, A. C. 1961. Growth from the vapor of large single crystals of lead selenide of controlled composition. *Journal of the Electrochemical Society* 108: 82–87.
17. Cutter, J. R., and J. Woods. 1979. Growth of single crystals of zinc selenide from the vapour phase. *Journal of Crystal Growth* 47: 405–413.
18. Glass, R. C., D. Henshall, V. F. Tsvetkov, and C. H. Carter. 1997. SiC-seeded crystal growth. *Materials Research Society Bulletin* 22: 33–35.
19. Acheson, A. G. 1892. British Patent No. 17, 911.
20. Lely, A. J. 1955. Darstellung von Einkristallen von SiliciumCarbid und Beherrschung von Art und Menge der einge bauten Verun reinigungen. *Berichte der Deutschen Keramischen Gesellschaft* 32: 229–236.
21. Tairov, Y. M., and V. F. Tsvetkov. 1978. Investigation of growth processes of ingots of silicon carbide single crystals. *Journal of Crystal Growth* 43: 209–212.
22. Barrett, D. L., R. G. Seidensticker, W. Gaida, and R. H. Hopkins. 1991. SiC boule growth by sublimation vapor transport. *Journal of Crystal Growth* 109: 17–23.
23. Dhanaraj, G., B. Raghoothamachar, and M. Dudley. 2010. Growth and characterization of silicon carbide crystals. In *Springer handbook of crystal growth*, edited by G. Dhanaraj et al. New York, NY: Springer, pp. 797–820.
24. Mahajan, S., M. V. Rokade, S. T. Ali, N. R. Munirathnam, S. Deb, D. V. Sidhararao, M. Vijayakumar, and A. K. Garg. 2014. *Growth of 2" silicon carbide (SiC) single crystal.* A paper presented at the 25th Annual Meetings of the Materials Research Society-India.
25. Piper, W. W., and S. J. Polich. 1961. Vapor phase growth of single crystals of II-VI compounds. *Journal of Applied Physics* 32: 1278–1279.
26. Liu, J. Z., J. W. Dykes, M. D. Lan, P. Klavins, R. N. Shelton, and M. M. Olmstead. 1993. Vapor transport growth of C_{60} crystals *Applied Physics Letters* 62: 531–532.
27. van Arekel, A. E., and J. H. de Boer. 1925. Darstellung von reinem Titanium-, Zirkonium-, Hafnium- und Thorium metal. *Zeitschrift fur Anorganische und Allgemeine Chemie* 148: 345–350.
28. Moliere, K., and D. Wagner. 1959. Herstellung von Einkristallen hochschmelzender. Metalle durch thermische Zersetzung von Halogeniddampfen. *Zeitschrift für Elektrochemie* 61: 65.
29. Webb, W. W. 1965. Dislocation mechanisms in the growth of palladium whisker crystals. *Journal of Applied Physics* 36: 214–221.
30. Schaffer, P. S. 1965. Vapor-phase growth of alpha alumina single crystals. *Journal of the American Ceramics Society* 48: 508–511.

11

Growth of Thin Films

Introduction

A monograph on crystal growth would be incomplete without a brief discussion on the growth of oriented thin films. As mentioned in the previous chapter, vapor phase techniques have been extensively used to grow epitaxial films. However, in applications requiring highly doped epitaxial films, such as tunnel and laser diodes, liquid phase epitaxy has advantages over vapor phase epitaxy. Further, there is also solid phase epitaxy, which occurs when a metastable amorphous layer is in contact with a single crystal template. In this chapter, we describe techniques available for the growth of thin films from vapor, liquid, and solid phases. This subject is vast, and scores of books have been written on it; a few are listed at the end of this chapter [1–4]. More recently, the *Handbook of Crystal Growth*, edited by Hurle, contains two volumes exclusively for thin films [5]. The entire part E of the *Springer Handbook of Crystal Growth* [6] is devoted to epitaxial growth and thin films. Readers may consult these books for the latest developments in the field.

Epitaxial Growth

A thin film may be defined as the one whose physical properties depend on the thickness. It is a layer of material ranging from a fraction of a nanometer (monolayer) to several micrometers in thickness. Since thin films are fragile and not self-supporting, they are invariably deposited on substrates. Oriented growth of films can occur on a variety of substrates, and the process is referred to as "epitaxial growth." Three parameters that influence the growth of epitaxial films are (1) evaporation rate, (2) temperature of the substrate, and (3) nature of the substrate.

When the evaporation rate is extremely high, there is only a limited time for the atoms/molecules on the surface to rearrange and combine to form a cluster called an island. Hence, rapidly deposited films initially contain a large number of small islands, whereas slowly deposited films contain a few large islands. As the average thickness increases, the distance between the islands decreases and eventually a continuous film is produced if the growth is allowed to continue for a sufficient period. Films grown with fast evaporation may not have smooth surfaces.

An important factor in the growth of the films is the surface mobility of the condensing species, and this ability is derived from the energy of condensation. If the substrate is at an elevated temperature, surface-adsorbed species have greater mobility, which in turn facilitates orderly growth leading to crystalline film. For obtaining oriented films, substrate of the same material with specific orientation is ideal, and the growth is known as homoepitaxy. Alternatively, lattice-matched substrates of different materials can be used, and the resulting growth is known as heteroepitaxy. Epitaxial films can also be grown on lattice-mismatched substrates, provided the mismatch is not too severe. In conventional lattice-matched epitaxy, if the lattice misfit is less than 7%–8%, the thin films usually grow pseudomorphically, but if it is above this, they will grow either textured or largely polycrystalline.

The lattice mismatch f is defined as

$$f = (a - a_0)/a_0 \tag{11.1}$$

where a is the lattice constant of the epilayer and a_0 is that of the substrate. On mismatched substrate the growing film initially is forced to take the lattice constant of the substrate, leading to a strained epilayer. However, as the thickness of the film increases, the strain relaxes through the formation of misfit dislocations, as shown in Figure 11.1. This happens because when the film is thin, the strain energy is lower, and when it becomes thick, the dislocation energy is lower. The thickness at which the two energies balance is known as the critical thickness and is approximately given by $(a/2f)$.

It goes without saying that the substrates on which the films are to be grown should be clean. The cleaning of the substrate prior to deposition may

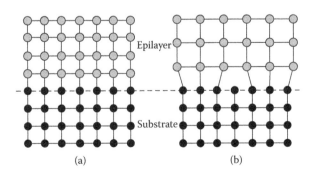

(a)   (b)

FIGURE 11.1
Growth on (a) lattice-matched and (b) lattice-mismatched substrates.

be performed in many ways. A simple procedure commonly used includes ultrasonic cleaning in detergent solution, followed by a thorough rinsing in deionized water, and finally rinsing with double-distilled acetone.

Growth of Thin Films from the Vapor Phase

Many physical vapor deposition techniques have been proposed for growing thin films. We begin with the simplest and the most commonly used technique (i.e., vacuum deposition) and then discuss the others.

Vacuum Deposition

This is a very versatile technique and can be used to deposit any material that either sublimes or melts congruently with sufficient vapor pressure. Materials having sufficient vapor pressure below the melting point can also be deposited. The equipment used for the vacuum deposition of thin films consists of two parts: (1) the vacuum chamber or bell jar with its associated pump, valves, base plate, and gauges and (2) the mechanical hardware installed within the chamber such as sources, masks, substrates, heaters, and thickness-monitoring equipment. The vacuum chamber is a glass jar, typically 30 inches high and 18 inches in diameter, but the dimensions vary depending on the need. A variety of evaporation sources, masks, and substrates are located inside the chamber. The evaporation sources are heated to temperatures as high as 3000°C. The atoms emitted from a given source propagate in straight lines until they strike the substrate through the appropriate mask with which it is in contact. By changing mask patterns and source materials, several films may be successively deposited on the substrate. Since the mean free path of the subliming species is inversely proportional to the pressure, better material transport is achieved if the sublimation occurs in high vacuum. Most commercial deposits are made within the 10^{-5} to 10^{-6} Torr region and rarely in the ultra-high vacuum (UHV) region. Figure 11.2 shows the arrangement used to deposit a single film on a heated substrate.

The selection of a good vapor source is crucial in making good thin films. The simplest vapor source is a hot wire, and it can be used when evaporation is by sublimation. This method is used to grow metals like tungsten and rhodium. If the material to be evaporated can be electroplated onto the wire, for example, chromium on tungsten wire, it can be sublimed off or melted to form drops that will then vaporize. Sources in the form of boats fabricated from the refractory metals are useful if a larger amount of vapor is needed. Metallic crucibles are usually used to vaporize dielectrics such as oxides of silicon, zinc, and so on. If the charge is available as a powder or mixture of powders, it can be dropped onto a hot ribbon and flash evaporated.

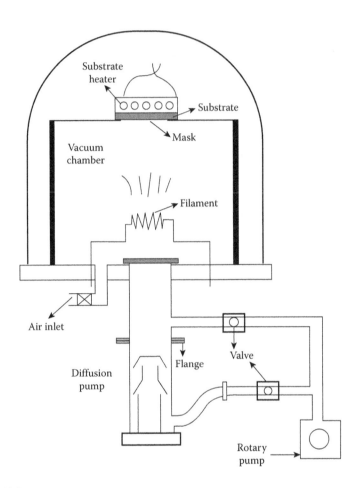

FIGURE 11.2
Conventional vacuum evaporation unit.

The best method for heating, however, is the electron beam heating, and the method is called electron beam evaporation.

One of the disadvantages of vacuum evaporation technique is the non-uniformity of film thickness, particularly when the films are large. Since the vapor species originate from point-like sources, they travel radially. Consequently, their travel time to reach the substrate kept above the source varies. This results in variation in film thickness laterally with reduced thickness toward the periphery.

Sputtering

If the positive ions in a gaseous discharge bombard a metallic cathode at sufficiently low gas (usually argon) pressures (10^{-2} to 10^{-1} Torr), free atoms leave the cathode and can be deposited on a surface kept in its vicinity.

This transfer of material from the cathode is known as cathode sputtering. The principal advantage of the sputtering technique is that the films are grown at a lower temperature than those grown by normal sublimation–condensation process. Also, because the target is an extended source (usually in the form of a circular disc), large films can be grown with uniform thickness. Figure 11.3 schematically shows a cathode, sputtering apparatus. Here, the material to be deposited is taken as the cathode, and the substrate is taken as the anode. A direct current (DC) potential of several kilovolts is applied between the anode and cathode. Both anode and cathode may be maintained at desired temperatures by independently controlling the heater inputs. Using suitable masks, films may be deposited over a required area and shape.

The sputtering rate to a good approximation is linearly proportional to the number and energy of the bombarding ions. It increases with the temperature

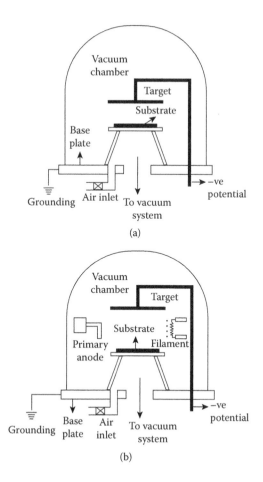

FIGURE 11.3
Schematic of (a) simple cathode sputtering apparatus and (b) hot cathode sputtering unit.

TABLE 11.1

Comparison between Vacuum Deposition and Sputtering
Techniques

Vacuum Evaporation	Sputtering
10^{-5}–10^{-6} Torr	10^{-1}–10^{-2} Torr
Variable thickness	Uniform thickness
Purer samples	Incorporation of sputtering gas possible
Small substrates	Large substrates
Only residual gas	Argon gas
Sublimation	Physical removal by momentum transfer

of both the discharge gas and the cathode and with the atomic weight of the discharge gas. By adding small amounts of other gases to the argon, the composition can be modified. This is known as reactive sputtering.

In hot cathode sputtering, a variant of sputtering, a filament serves as a hot cathode, which when energized, emits electrons to ionize the sputtering gas (Figure 11.3b). An ion target or source material is introduced as a third element and with a negative potential attracts the positive ion in the argon gas. The use of the thermionic cathode permits a glow discharge to be maintained at a pressure as low as 1×10^{-4} Torr, which is two orders of magnitude lower than that of normal cathode sputtering. This method therefore permits higher deposition rates when compared with cathode sputtering.

Magnetron sputtering, yet another variant of the sputtering technique, uses the effect of the magnetic field on the bombarding electrons advantageously. The effect of the magnetic field in a sputtering configuration is the confinement of electrons on the target surface (cyclotron effect). This increases ionization; hence, the sputtering rate increases.

Table 11.1 compares the two methods prominently used for producing thin films just described. A very significant difference exists in the films produced by these techniques due to the different vacuum conditions.

Pulsed Laser Deposition

The availability of high-energy pulsed lasers has led to the development of a new thin film preparation technique called laser ablation or more popularly pulsed laser deposition (PLD) technique. The real breakthrough for the PLD technique came in 1987 when Dijkkamp, Venkatesan, and coworkers [7] deposited superior quality thin films of $YBa_2Cu_3O_7$ (a high-temperature superconductor) using this technique compared to those deposited by alternative techniques. Since then, the technique of PLD has been utilized to fabricate high-quality crystalline films of ceramic oxides, nitrides, metallic multilayers, and various superlattices.

In this process, a laser source kept outside a vacuum chamber sends high-energy pulses into the chamber through a proper window, where it is

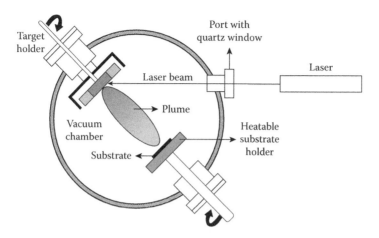

FIGURE 11.4
Schematic of pulsed laser deposition setup.

focused on the target at a 45° incident angle. The impact of the laser results in the ablation of the target material from its surface. In fact, the impact leads to many processes, including ablation, melting, and evaporation, as well as the creation of plasma due to ionization of the ejected species forming a plume. The material ejected from the target is deposited onto a substrate, which is kept just opposite to it, as shown in Figure 11.4.

The laser that is most commonly used is the pulsed excimer laser in the ultraviolet range (usually 193 or 248 nm). Other lasers like ruby and Nd–YAG have also been used for the growth.

Apart from enabling the growth of multicomponent systems, one of the major advantages of the PLD technique is that it facilitates the growth of films with the same composition as that of the target. Also the deposition rates are relatively high. Further films with desired microstructures can be obtained at relatively low substrate temperatures. However, PLD also has its disadvantages, the main among them being the particle formation during ablation. These particles get incorporated into the growing film, thereby affecting its quality. Also, the scaling up of the PLD technique to prepare large area films is rather difficult because of the focused nature of the ablated plume.

Chemical Vapor Deposition

In the previous chapter, we saw how chemical vapor transport (CVT) technique is used for the growth of bulk single crystals. This technique can readily be adapted for the deposition of thin films and is called chemical vapor deposition (CVD) technique. There are three different CVD versions used for the growth of semiconductors. They are (1) the halide process, (2) the hydride process, and (3) the metal-organic chemical vapor deposition process.

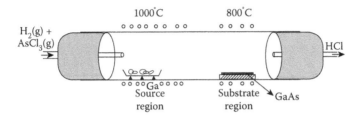

FIGURE 11.5
Typical CVD process to grow GaAs.

Of these the first one is nothing but the CVT with the aid of halogens as a transporting agent.

The halide process has been used for the growth of GaAs. Here, the source materials are high-purity elemental Ga and $AsCl_3$, a liquid. The $AsCl_3$ is kept in bubblers through which hydrogen is passed. The resulting gas flows over the Ga source and then onto the substrate. The reactor tube is heated at the source region (boat containing Ga) to about 1000°C and the substrate region to about 800°C–850°C (Figure 11.5).

When H_2 saturated with $AsCl_3$ flows over the source region, the following reaction takes place:

$$2AsCl_3(g) + 3H_2(g) \rightarrow 2As(g) + 6HCl(g) \qquad (11.2)$$

The HCl formed in this reaction further reacts with the molten Ga to produce GaCl as follows:

$$2Ga(l) + 2HCl \rightarrow 2GaCl(g) + H_2(g) \qquad (11.3)$$

The following reaction then takes place over the heated substrate, leading to the growth of GaAs:

$$2GaCl(g) + 2As(g) + H_2(g) \rightleftharpoons 2GaAs(s) + 2HCl(g) \qquad (11.4)$$

The reaction is exothermic and reversible to a certain extent. Usually, Zn is used for p-type doping and S/Se in the form of H_2S/H_2Se for n-type doping. Since it is possible to obtain high-purity liquid sources through the distillation process, chloride-based CVD produces very pure epilayers.

In the hydride process, gaseous AsH_3 or PH_3 is used as a source for the group V elements. The hydrides decompose in the high-temperature region of the furnace as follows:

$$2AsH_3(g) \rightarrow 2As(g) + 3H_2(g) \qquad (11.5)$$

Hydrochloric acid reacts with the metallic group III source to produce metallic chloride, according to Equation 11.3. Finally in the substrate region, epitaxial growth occurs as per Equation 11.4. The hydride process gives better control over the ratio of group III to group IV vapor phase species. It is the most common technique used to manufacture LEDs and detectors.

Metal-Organic Chemical Vapor Deposition

One of the techniques, frequently used for the fabrication of semiconductor devices, is metal-organic chemical vapor deposition (MOCVD). The method uses metal alkyls and hydrides as source materials in a cold-wall reactor. Industrial reactors have the capability of simultaneous deposition on as many as 50 wafers of a 50 mm diameter. The application of this technique has been extended to oxides, metals, and even organic materials. A number of advantages exist for the commercial use of MOCVD compared to other vapor phase techniques. First, very sharp changes in the compositions of the growing layers are possible with MOCVD because the reactions occurring here are not reversible. Second, the crystal growth occurs at relatively low temperatures, which minimizes the effect of inter diffusion. Third, phosphorous does not pose special problem in MOCVD, unlike in molecular beam epitaxy growth, which is discussed later.

The MOCVD was pioneered by Manasevit, who along with Simpson [8] reported the growth of GaAs on GaAs and several other substrates. Growth was achieved by pyrolyzing a mixture of trimethyl gallium (TMGa) and arsine (AsH_3) at a temperature between 600°C and 700°C in H_2 atmosphere. A representative list of metal alkyls is given in Table 11.2. They are all volatile liquids with the exception of TMIn, which is a waxy clear solid at room temperature. In general, lighter methyl compounds have higher vapor pressures.

The schematic of a typical vertical atmospheric pressure MOCVD reactor arrangement that can be used to grow $Al_xGa_{1-x}As$ layers is shown in Figure 11.6. It consists of an open-tube reactor containing a conducting graphite susceptor, which is heated either by RF induction or IR lamps. Substrates are placed on the susceptor. Dopants flow as diethyl zinc and the hydride of Se here. The other types of reactors in use are horizontal reactor, high-capacity barrel reactor, and the rotating disc reactors.

TABLE 11.2

Representative Metal Alkyls Used in MOCVD

Compound	Acronym	Melting Point (°C)	Boiling Point at 760 Torr (°C)	Vapor Pressure Torr
Trimethyl gallium	TMGa	−15.8	55.7	64.5 at 0°C
Triethyl gallium	TEGa	−82.3	143	18 at 48°C
Trimethyl aluminum	TMAl	15.4	126	8.4 at 20°C
Trimethyl indium	TMIn	88.4	134	1.7 at 20°C
Triethyl indium	TEIn	−32	184	3 at 53°C
Diethyl zinc	DEZn	−28	117	16 at 25°C

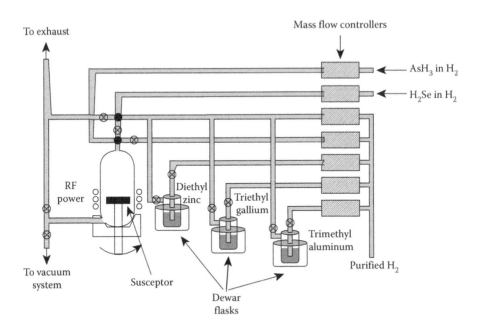

FIGURE 11.6
Schematic of an MOCVD setup.

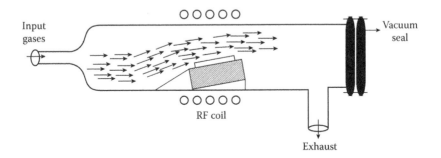

FIGURE 11.7
Horizontal reactor.

Of these, the horizontal geometry (Figure 11.7) is often preferred because gas flow can be more or less linear here. The susceptor in horizontal systems is often inclined at several degrees from the horizontal (15°–30°) to overcome the reactant depletion effects. Because inclination alone may not completely compensate for the reactant depletion, substrate rotation has also been applied.

In a typical growth run, TMGa and trimethyl aluminum (TMAl) are fed to the growth tube as gas flows by bubbling H_2 through bubblers containing liquid TMGa and TMAl. These are mixed with AsH_3 and pyrolyzed above the substrate. Gas flows are usually controlled by mass flow controllers. The net reaction for the growth of $Al_xGa_{1-x}As$ is given by

$$(1-x)[(CH_3)_3Ga] + x[(CH_3)_3Al] + AsH_3 \xrightarrow[700°C]{H_2} Al_xGa_{1-x}As + 3CH_4 \quad (11.6)$$

The alloy composition is directly determined by the relative initial partial pressures of TMGa and TMAl. In the case of horizontal reactors operating at low pressures, there is evidence for the existence of a so-called boundary layer through which the gas molecules have to diffuse to reach the substrate. Several groups working on GaAs have established that this process is the rate-limiting step.

Growth of Diamond Films by the CVD Process

Since the early 1980s, the CVD method has also been used to grow diamond films. The advantages of the CVD process to grow diamond include the ability to deposit diamond over large areas and on various substrates, and also the fine control one can exercise over the chemical impurities that are introduced during growth.

For single crystal growth, the substrate needs to be a single crystal diamond and the diamond film grows epitaxially on it. For the growth of polycrystalline diamond, a variety of materials such as silicon, silicon carbide, and a range of carbide-forming metals, including molybdenum and tungsten, have been used as substrate. The substrate is usually maintained at around 800°C. The growth process generally takes place at a pressure between 1 and 200 Torr. The precursors for diamond synthesis are hydrogen and a hydrocarbon such as methane, which are introduced into the growth chamber in a controlled manner. Only a tiny amount of hydrocarbon, mostly between 1% and 5%, is normally present in the gas mixture and provides the source of carbon from which the diamond is formed. Hydrogen, which is the major component of the gas mixture, should be present in the form of a hydrogen radical, if the process has to work. Therefore, gases are ionized into chemically active radicals in the growth chamber. This is achieved using microwave power, a hot filament, an arc discharge, a welding torch, a laser, or an electron beam.

Under the conditions used for the growth of CVD diamond, graphite is the thermodynamically favored phase rather than diamond. However, hydrogen radicals etch away any graphite forming on the substrate far faster (10–100 times faster) than they can remove any diamond. Thus, the kinetics of the process at the surface ensures that only diamond is left behind at the end of the growth run. Hydrogen also helps to terminate the dangling carbon bonds at the growing diamond surface, thereby preventing them from reconstructing into a nondiamond form.

Molecular Beam Epitaxy

Molecular beam epitaxy (MBE) is a process of growing epitaxial films on a heated substrate under UHV conditions using molecular or atomic beams. The beams are thermally generated in Knudsen-type effusion cells, which contain the constituent elements or compounds required to grow the epitaxial film. The cells are maintained at the required temperature accurately to give the thermal beams of appropriate intensity. The beams escaping from the orifices in the cells travel in rectilinear paths to the substrate, where they condense and grow under controlled conditions.

Single crystalline growth of GaAs films was first reported by Arthur [9] in 1968. However, since then, tremendous evolution has taken place in MBE growth and device quality. GaAs, AlGaAs films, semiconductor superlattices, quantum well structures, and so on have been grown. In recent years, molecular beam epitaxy has also been used to deposit oxide materials suitable for electronic, magnetic, and optical applications. MBE systems have been modified to incorporate oxygen sources to grow these oxide films [10].

The basic MBE process can be understood with the help of the schematic diagram shown in Figure 11.8. The molecular beams are generated in Knudsen-type effusion cells whose temperatures are controlled to ±1°C accuracy, in a UHV system. The cells are made from nonreactive refractory materials like pyrolytic boron nitride or high-purity graphite that can withstand the operating temperature. The various cells are placed in such a way that their beams converge on the substrate for epitaxial growth. Individual cells are surrounded by liquid nitrogen shrouds to prevent cross-heating

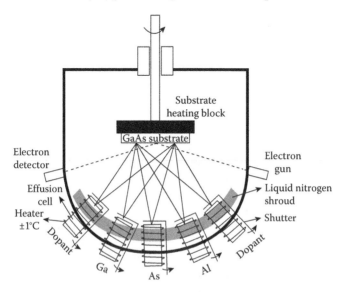

FIGURE 11.8
Schematic of a molecular beam epitaxy unit.

and cross-contamination. Mechanical shutters are placed in front of each cell and can be operated externally. This facilitates rapid changing of the beam species so that the composition and/or doping of the growing films can be abruptly altered.

The substrate holder is usually made out of Mo, on which the oriented substrate is mounted using indium. A resistance heater kept behind the Mo block is employed to heat the substrate to the growth temperature. The substrate holder can be rotated during deposition to achieve uniform epitaxial layers. Rotation speeds ranging from 0–30 rpm are available in commercial units. A sample exchange load-lock system is usually incorporated in the MBE system to decrease the time interval between successive growth runs.

A unique feature of MBE is that the UHV condition allows incorporation of surface analytical tools, such as Auger electron spectroscopy (AES) and reflection high-energy electron diffraction (RHEED). A quadrupole mass spectrometer can also be used to detect the residual gas composition in the chamber. While AES is used to characterize the initial surface composition, the RHEED provides information on surface reconstruction, microstructures, and smoothness.

Although conventional MBE has been very successful in the growth of many compound semiconductors, growth of III-IV compounds containing P is extremely difficult with it since elemental solid phosphorous consists of allotropic forms with different vapor pressures (P_4 and P_2). To avoid this, Panish and his group [11,12] devised a method of decomposing AsH_3 and PH_3 to obtain As_2 and P_2 for MBE growth of GaAs and InP, respectively. This MBE process has come to be known as gas source molecular beam epitaxy (GSMBE). High-resolution lattices imaging technique has shown that films as thin as three molecular layers can be grown using gas sources.

Liquid Phase Epitaxy

Liquid phase epitaxy (LPE) is the simplest and the most widely used technique for growing epitaxial layers of many group III–V compounds. This technique was first introduced by Nelson [13] in 1963 to grow GaAs laser diode structures. Since then, the method has been widely practiced and applied to grow not only III–V compounds but also II–VI compounds, carbides, nitrides, and a variety of oxides, including high T_c superconductors.

Basically LPE involves the growth of an epitaxial layer on a single crystal substrate from a solution supersaturated with the material to be grown. The substrate usually has a similar crystal structure and lattice constant of the growing layer to allow epitaxial growth. The growth solution is rich in one of the major components of the solids and dilute in all others. Compared to vapor phase epitaxy, the growth rates are high in LPE.

The thermodynamic basis of LPE can be illustrated by the phase diagrams shown in Figure 11.9. Referring to Figure 11.9a, when a liquid solution of composition X_2 at temperature T_2 is cooled to temperature T_1, the deposition

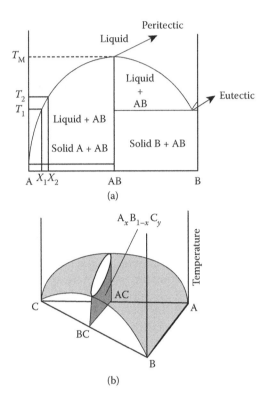

FIGURE 11.9
Typical (a) binary and (b) ternary phase diagrams.

of an amount of solid AB onto the substrate equivalent to the loss of $(X_2 - X_1)$ atom fraction of B and an equal amount of A would occur. The growth of ternary compounds $A_xB_{1-x}C$ can also be understood similarly with the help of a ternary phase diagram. Here, A and B are group III elements and C is a group V element. The composition of liquid ABC at a given temperature can be completely represented by the liquidus curve and the corresponding tie lines (Figure 11.9b).

In the case of LPE, kinetics of growth has significant influence on the formation of the film. Arrival rate of solute atoms at the growing interface, convection due to compositional and temperature gradients, imposed supersaturation, and the operating growth mechanism all influence the LPE growth rates.

Significant advantages of LPE as a relatively low temperature process can be understood with reference to the schematic phase diagram of GaAs shown in Figure 11.10 [14]. As can be inferred from the figure, since congruent composition of GaAs is off-stoichiometric, crystal grown at its melting temperature (1238°C) will have Ga deficiency and hence inhomogeneities in Ga carrier concentrations. The temperature range used for the LPE growth,

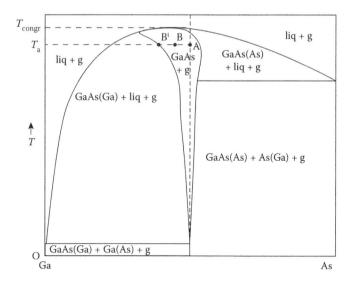

FIGURE 11.10
Schematic of the phase diagram of GaAs. (Reprinted from *Journal of Physics and Chemistry of Solids*, 32, Logan, R. M., and Hurle, D. T. J., Calculation of point defect concentrations and non-stoichiometry in GaAs, 1739–1753, 1971, with permission from Elsevier.)

on the other hand, is 700°C–900°C, which is much lower than the melting point and consequently with less defect concentration. This has been found to be the most desirable temperature range for growing stoichiometric films.

Three basic techniques are used for growing thin films by LPE: (1) tipping, (2) dipping, (3) and boat sliding. These along with the procedures are described next.

Tipping Technique

Figure 11.11a illustrates the cross-sectional view of the apparatus used to grow GaAs film by the tipping technique. The substrate is held tightly against the flat bottom at the upper part of the graphite boat, which is fixed in the center of a quartz tube furnace where the temperature is uniform. A mixture of tin and GaAs in required proportion is placed at the lower end of a graphite boat. The furnace tube is initially tipped, as shown in the diagram, and heated to about 840°C with hydrogen flowing. As the temperature rises, GaAs particles dissolve in tin at the lower end of the boat. When the temperature reaches about 840°C, the heating power is turned off and the furnace is tipped so that the molten tin nearly saturated with GaAs covers the exposed surface of the GaAs wafer. As the furnace cools, GaAs initially dissolves from the substrate surface to reach equilibrium, thereby cleaning the interface. On further cooling, the solution gets supersaturated with GaAs and epitaxial growth of GaAs occurs on the substrate.

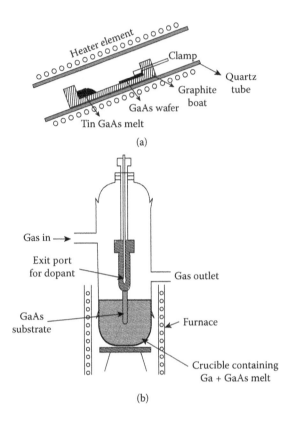

FIGURE 11.11
(a) Tipping technique and (b) dipping technique.

When the temperature reaches ~600°C, the solution is decanted from the wafer surface by tipping the furnace back to its original position. The flow of hydrogen is replaced by nitrogen, and the graphite boat is removed from the furnace tube. Any remaining tin on the surface of the epitaxial layers is wiped off before it solidifies. The same solution can be used successively for processing several wafers. Occasionally, GaAs is added to the melt to compensate for that consumed in epitaxial growth.

Dipping Technique

The dipping technique uses a vertical furnace, as shown in Figure 11.11b. Here, the solution is contained in a graphite or alumina crucible at the lower end of the furnace, and the substrate fixed in a movable holder is initially positioned above the solution. At the desired temperature, the growth is initiated by immersing the substrate in the solution and is terminated by its withdrawal. The vertical dipping technique can also be used for multilayer growth in which the graphite boat containing several bins is rotated to transport the solution to the position below the seed for successive dipping.

Sliding Boat Technique

This technique uses the multibin boat to grow multiple epitaxial layers. The principal components of this apparatus are massive graphite boat with many bins and a graphite slide, a fused silica growth tube to provide a protective atmosphere, and a horizontal tubular furnace. The schematic of the multibin boat and a photograph of the boat are shown in Figure 11.12.

The graphite boat has the desired number of solution bins, depending on the number of layers to be grown, and the slider has two substrate slots, one for the sacrificial substrate and the other for the growth substrate. More slots can also be provided to place many substrates. To further improve throughput,

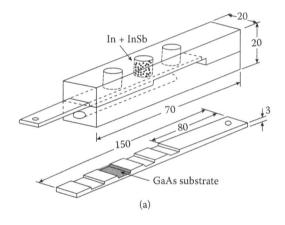

(a)

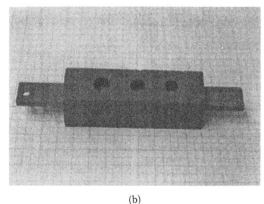

(b)

FIGURE 11.12
Multibin boat and the slider (a) schematic and (b) photograph.

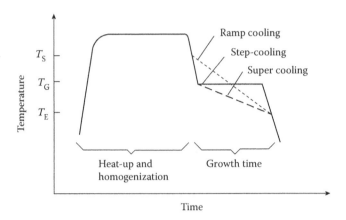

FIGURE 11.13
Thermal cycle used in the sliding boat technique.

a multisubstrate LPE with complicated boat designs has been developed in which substrates sliders are stacked vertically in different cavities. The function of the sacrificial substrate is to saturate the growth solution before it is brought into contact with the growth substrate. The substrates are brought into contact with each solution by the relative motion of barrel over the slides, which are automated in commercial units. Growth is usually carried out in an atmosphere of H_2. An approximate temperature profile for the successive growth is usually predetermined.

Typical thermal cycling applied to the system is shown in Figure 11.13. Initially, the system is heated up to a temperature slightly above T_s in order to homogenize the solution. It is then cooled down to the starting temperature T_s, during which the substrate is brought into contact with the solution. The growth is finally terminated at the end temperature T_e. Three different types of cooling are normally employed, that is, step cooling, ramp cooling, and supercooling, as shown in Figure 11.13.

One of the problems encountered with the sliding boat LPE growth technique is the carryover of one solution into the next. Solution carryover can cause variation in layer thickness, composition, interface properties, and dopant profiles. Since the amount of solution carryover increases with increasing clearance between the substrate and the walls of the slit, minimizing slide clearance can minimize carryover. However, minimizing the clearance has the adverse effect of scratching the grown layer; hence, a compromise must be made.

Surface Morphology

One of the limitations of LPE growth is the difficulty in achieving a flat and featureless surface of the grown layers. There are three distinct types of

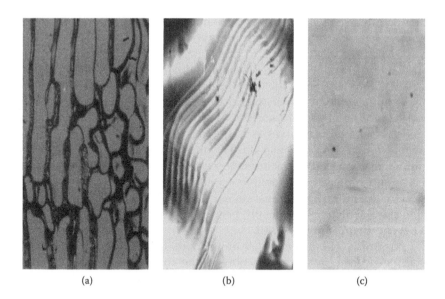

(a) (b) (c)

FIGURE 11.14
Surface features on LPE-grown epilayers. (a) Island growth (GaSb), (b) terracing (InSb), and (c) almost featureless surface. Faint meniscus lines can be seen here (InSb). (Panel a is reprinted from *Bulletin of Materials Science*, 18, Dutta, P. S., et al., Liquid phase epitaxial growth of pure and doped GaSb layers: morphological evolution and native defects, 865–874, 1995, with permission from Elsevier. Panels b and c are from Udayashankar, N. K., Growth and characterization of certain Sb-based III-V semiconductors, PhD dissertation, Indian Institute of Science, Bangalore, India, 1999. With permission.)

surface features associated with the LPE layers. These are growth islands, terracing, and meniscus lines. Island growth arises as a result of inadequate nucleation due to wetting problem, surface oxidation, contamination, and so on. Terracing is observed where layers are grown on substrates whose orientations are close to low-index planes. Such misoriented substrates have steps of microscopic dimension. During growth, the step-bunching process occurs, leading to terraces. Meniscus lines are formed as a result of the motion of trailing edges of liquid as it is moved across the crystal during the sliding procedure. Figure 11.14 shows some of the surface features observed on LPE-grown GaSb [15] and InSb [16] layers.

Solid Phase Epitaxy

We conclude this chapter with a brief note on solid phase epitaxy. Solid phase epitaxy is a crystal growth process in which an amorphous phase of a material transforms into its crystalline phase due to reconstruction. This was first reported in silicon [17] and gallium arsenide [18], which led to extensive studies on other semiconducting compounds. However, silicon still remains the most extensively studied electronic material.

In solid phase epitaxy, the amorphous phase is metastable, and the transformation to crystalline state is associated with an abrupt change in the free energy of the system. Since the transformation is purely a solid–solid transition, the atomic mobility is relatively low in comparison to that occurring in vapor–solid or liquid–solid transition. The ordered structure on the crystalline side of the interface serves as a substrate for the layer-by-layer addition of atoms from the amorphous side. The amorphous layers on the substrate can either be deposited by vacuum deposition or prepared by ion implantation. The substrate is then heated to crystallize the film. During the process of growth, the impurity segregation and redistribution occurs at the growing interface, which is beneficially used to incorporate low-solubility dopants beyond their solubility limit.

The possibility of converting an amorphous layer into a highly ordered crystalline layer has many applications in the area of thin films technology. Recovery of damage caused by ion implantation, epitaxial growth on highly lattice-mismatched substrates, and epitaxial growth of excessively doped semiconductor thin films are some examples.

References

1. Holland, H. 1956. *Vacuum deposition of thin films*. London, UK: Chapman and Hall.
2. Chopra, K. L. 1969. *Thin film phenomena*. New York, NY: McGraw Hill.
3. Maissel, L. I., and R. Glang, eds. 1970. *Handbook of thin film technology*. New York, NY: McGraw Hill.
4. Mathews, J. W., ed. 1975. *Epitaxial growth*. New York, NY: Academic Press.
5. Hurle, D. T. J., ed. 1994. Thin films and epitaxy. In *Handbook of crystal growth*, Vols. 3a–b. Amsterdam, The Netherlands: North Holland, pp. 1–1047.
6. Dhanaraj, G., K. Byrappa, V. Prasad, and M. Dudley, eds. 2010. *Springer handbook of crystal growth*. New York, NY: Springer, pp. 939–1211.
7. Dijkkamp, D., T. Venkatesan, X. D. Wu, S. A. Shaheen, N. Jisrawi, Y. H. Min-Lee, W. L. McLean, and M. Croft. 1987. Preparation of Y·Ba·Cu oxide superconductor thin films using pulsed laser evaporation from high T_c bulk material. *Applied Physics Letters* 51: 619–621.
8. Manasevit, H. M., and W. I. Simpson. 1969. The use of metal-organics in the preparation of semiconductor materials: I. Epitaxial gallium-V compounds. *Journal of the Electrochemical Society* 116: 1725–1732.
9. Arthur, J. R. 1968. Interaction of Ga and As_2 molecular beams with GaAs surfaces. *Journal of Applied Physics* 39: 4032–4034.
10. Cheng, J., V. K. Lazarov, G. E. Sterbinsky, B. W. Wessels, and V. Lazarov. 2009. Synthesis, structural and magnetic properties of epitaxial thin films by molecular beam epitaxy. *Journal of Vacuum Science and Technology B* 27: 148–151.

11. Panish, M. B. 1980. Molecular beam epitaxy of GaAs and InP with gas sources for as and P. *Journal of the Electrochemical Society* 127: 2729–2733.
12. Panish, M. B., H. Temkin, and S. Susuki. 1985. Gas source MBE of InP and $Ga_x In_{1-x} P_y As_{1-y}$: Materials properties and heterostructure lasers. *Journal of Vacuum Science and Technology B* 3: 657–665.
13. Nelson, H. 1963. Epitaxial growth from the liquid state and its application to the fabrication of tunnel and laser diodes. *RCA Review* 24: 603–615.
14. Logan, R. M., and D. T. J. Hurle. 1971. Calculation of point defect concentrations and nonstoichiometry in GaAs. *Journal of Physics and Chemistry of Solids* 32: 1739–1753.
15. Dutta, P. S., H. L. Bhat, and V. Kumar. 1995. Liquid phase epitaxial growth of pure and doped GaSb layers: Morphological evolution and native defects. *Bulletin of Materials Science* 18: 865–874.
16. Udayashankar, N. K. 1999. Growth and characterization of certain Sb-based III-V semiconductors. PhD dissertation, Indian Institute of Science, Bangalore, India.
17. Mayer, J. W., L. Eriksson, S. T. Picraux, and J. A. Davies. 1968. Ion implantation of silicon and germanium at room temperature. Analysis by means of 1.0-MeV helium ion scattering. *Canadian Journal of Physics* 45: 663–673.
18. Cho, A. Y. 1969. Epitaxy by periodic annealing. *Surface Science* 17: 494–503.

12

Crystal Characterization

Introduction

It is well established that no crystal is perfect; hence, in crystals produced either naturally or artificially, one would always find deviations from the ideal structure. These deviations may be in terms of compositional inhomogeneities, impurities, or defects, and all of these influence most of the physical properties of the crystal. Even sophisticated measurements, when carried out on bad crystal, may give misleading results. Hence, characterization of a crystal is a prerequisite for a meaningful interpretation of the obtained results. It is also important from the application point of view.

The term *characterization* as such is too general. Also, complete characterization of a crystal is quite involved in terms of both instrumental requirement and time. It actually requires a group of specialists well versed in sophisticated techniques as well as in interpretation of the obtained results, which most of the crystal growth laboratories may not have. However, certain parameters should always be specified for a crystal. In this chapter, we deal with such characterization techniques that are required for specifying these parameters. The book *Crystal Growth from High Temperature Solution* by Elwell and Scheel [1] has a lengthy chapter on characterization that discusses the importance, complexities, and problems associated with this subject.

Nomenclature

Nomenclature is a system of names or terms used with which we identify and distinguish the objects that we encounter, along with their similarities and differences. It also helps us to possibly classify the objects of interest. In the context of crystals, identification is quite important so that we can avoid uncertainty and confusion. Unless the full name of the crystal along with its chemical formula and structure is given, in many cases we may not be able to identify the crystal. J. C. Brice in his book [2] drives home this point with two typical examples. The first one is that of a well-known material, zinc sulfide.

We know that this material occurs in two forms that are completely different even though their chemical formula is the same: ZnS. Unless one adds the prefix α or β, the information given would be incomplete. While α-ZnS is hexagonal with a wurtzite structure, the β variety is cubic with a sphalerite structure. Both varieties can be produced in the laboratory. Hence, full information is essential for identification.

Brice's second example is bismuth germanium oxide, where the chemical formula distinguishes one crystal from the other but not the name. Here, again the three varieties represented by the chemical formulae $Bi_{12}GeO_{20}$, $Bi_4Ge_3O_{12}$, and $Bi_2Ge_3O_9$ are distinctly different, in terms of both structure and physical properties, but they are probably called by the same name. In particular, $Bi_{12}GeO_{20}$ is a photorefractive crystal, while $Bi_4Ge_3O_{12}$ is a nuclear detector material. The compounds $KNbO_3$, $K_4Nb_6O_{17}$, $K_2Nb_4O_{11}$, and KNb_3O_8, all probably called by the same name, are other examples wherein the identification is through the formula and not by the name [3]. In literature, we may find many such examples where nomenclature would be extremely important to distinguish the crystal.

Along with the name and formula, if the method of growth is also indicated, it would suggest what could be the level of impurities and defect concentration present in the crystal. For example, if the crystal is grown by a crucibleless technique, there would be no container-related contamination. Further, if the crystal is grown by a low-temperature method, one can expect a lower concentration of thermally induced defects in it. Crystals grown under isothermal conditions have better compositional homogeneity than the ones grown in a thermal gradient. Thus, the method of growth provides useful qualitative information about the crystal. Size and shape of the crystal would also be useful inputs from the user's point of view.

Chemical Composition

The need to specify the chemical composition of the crystal arises because of the influence it has on the physical properties. First, we should know with sufficient accuracy the concentration of the major constituents so that deviation from stoichiometry, if any, may be estimated. Its importance can be illustrated with the help of a well-known crystal, $LiNbO_3$. As we saw earlier, the congruent melting composition for this crystal is not stoichiometric and the Li/Nb ratio influences its ferroelectric transition temperature T_c. For example, a crystal with 51% Li has 1200°C as its transition temperature, whereas for 47% Li concentration the T_c drops to 1060°C. Determination of concentrations of constituents is important in the case of compound semiconductors, solid solutions, alloys, multicomponent systems, and so on. However, this is not an issue with crystals with a single element. Second, if the crystal is intentionally doped, which is usually done to alter its physical properties, dopant concentration must be determined. This is extremely important because in many cases what has been added to the starting material would not get into the crystal. Third, we should be able to estimate the concentration of trace impurities incorporated inadvertently. The sources of these impurities may be the precursors themselves, crucible contamination, or even the imposed ambience, and they may be in minute quantities. Equally important to know is the inhomogeneity in their distribution. In compounds containing elements with multiple valance states, the determination of valance state of the constituent element also becomes important.

There are a number of physical and chemical techniques that may be employed to determine the chemical composition of crystals. Even though the classical method involving wet chemistry is still in use, this has been largely replaced by less laborious physical techniques. Next we discuss three most commonly used techniques for the determination of chemical composition.

CHN Analysis

CHN analysis is a form of elemental analysis concerned with determination of carbon, hydrogen, and nitrogen in a sample. In many cases, particularly when the sample is organic or semiorganic, this is the only investigation performed to estimate the elemental composition. Numerous organic compounds include no additional elements besides C, H, and N, except oxygen, which is seldom determined separately.

The technology behind the CHN analysis is combustion where the sample is first fully combusted and then the products of the combustion are analyzed. This is done by passing the final product (gas mixture) through a gas chromatographic system using high-purity helium as a carrier gas. Separation of the species is done by so-called zone chromatography. The measured weights are used to determine the elemental composition of the analyzed sample. The commercial instruments are usually calibrated with standard substances provided by the suppliers. The detection limit for carbon and nitrogen in ideal cases can be in the range of 500 ppm. Typical data would look like that shown in Table 12.1 [4].

TABLE 12.1

CHN Analysis of Sodium *p*-Nitrophenolate Dehydrate

	$Na(C_6H_4NO_3)2H_2O$		
	C(%)	H(%)	N(%)
Computed	36.50	4.10	7.10
Experimental	36.47	4.02	6.99

Source: Brahadeeswaran, S., et al., *Journal of Material Chemistry*, 8, 613–618, 1998.

Energy-Dispersive x-Ray Analysis

Energy-dispersive x-ray analysis (EDAX) is another widely used analytical technique for the elemental analysis of a sample. The facility is usually available with electron microscopes where high-energy electron beams are available. A typical EDAX setup consists of an excitation source (electron beam or x-ray beam), an x-ray detector, a pulse processor, and an analyzer.

The technique's characterization capability arises from the fact that each element has a distinct atomic structure, giving rise to a unique set of peaks in its x-ray spectrum when excited. To cause the emission of characteristic x-rays from a specimen, a high-energy electron (proton or x-ray) beam is focused onto the sample under investigation. The incident beam excites an electron in an inner shell, ejecting it from the shell, thereby creating an electron hole. Consequently, an electron from an outer, higher energy shell fills this hole, and the difference in energy between the two shells is released in the form of an x-ray. The number and energy of the x-ray photons emitted from a specimen are measured by an energy-dispersive spectrometer. As the energy of the x-rays is characteristic of the atomic structure of the element from which they were emitted, the elements of the specimen can be identified. The intensities (number of photons) of the x-rays emitted, on the other hand, give information about their concentration. Typical EDAX spectra are shown in Figure 12.1 [5], wherein a small amount of Bi added to $InAs_xSb_y$ shows up.

Often the excess energy of the electron that migrates to an inner shell to fill the newly created hole is transferred to a third electron from a further outer shell, prompting its ejection. This ejected species is called an Auger electron, and the method for its analysis is known as Auger electron spectroscopy. X-ray photoelectron spectroscopy is another variant of EDAX that utilizes the ejected electrons for analysis. Information on the number and

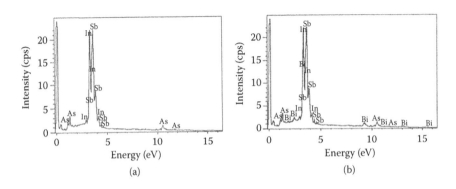

(a)   (b)

FIGURE 12.1
EDAX spectra of (a) In As_xSb_{1-x} and (b) In $Bi_xAs_ySb_{1-x-y}$. (From Dixit, V. K., Bulk and thin film growth of pure and substituted indium antimonide for infrared detector applications, PhD dissertation, Indian Institute of Science, Bangalore, India, 2003.)

kinetic energy of ejected electrons is used to determine the binding energy of the newly liberated electrons, which is element specific and allows chemical characterization of a sample.

With modern detectors and electronics, most EDAX systems can detect x-rays from all the elements in the periodic table above beryllium, provided they are in sufficient quantities. Even though a detection limit as low as 0.02% has been claimed for elements with an atomic number greater than 11, for routine measurement in an electron microscope the minimum detection limit is more like 1%–2%. This limit is essentially with respect to the major constituents, and for a general analysis, a more conservative figure of 5% is applicable.

Inductively Coupled Plasma Atomic Emission Spectroscopy

Inductively coupled plasma atomic emission spectroscopy (ICPAES), also referred to as inductively coupled plasma optical emission spectrometry, has become a very popular analytical technique since the advent of a modern excitation source (inductively coupled plasma) and improved electronic detection method. Concentrations of over 70 elements in the periodic table are routinely determined with varying detection limits (sub ppb to ppm levels).

The ICPAES setup consists of a sample introduction system, a plasma torch along with its power supply and impedance matcher, and an optical measurement system (which is usually a grating spectrometer), as shown schematically in Figure 12.2. A high-frequency induction heater is used to heat a stream of argon and form plasma via an induction coil. The temperature reached varies between 6000 and 8000 K. The sample must be introduced into the plasma as small droplets of solution (i.e., mist). When these atoms are excited by the plasma, they re-emit the energy acquired in the form of electromagnetic radiation. Usually UV and visible regions of the spectrum are recorded, as most of the intense emission lines are in between 200 and 400 nm. The emitted light beam is focused by a convergent

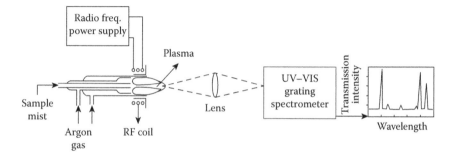

FIGURE 12.2
Schematic of an ICPAES.

lens onto the primary slit of a monochromator, which separates the beam into the component lines of different wavelengths corresponding to the elements present in the test sample. The intensity of light emitted at various wavelengths is related to the concentration of each element via calibration curves. The calibration curves are drawn by measuring the intensities of known concentrations of the elements. The relationship between emission intensity and concentration is known to be linear over five to six orders of magnitude.

Dopants and Impurities

The need to specify the dopant concentration is apparent. Concentrations of the intentionally doped elements define the physical properties that are intended to be modified by doping. For example, the dopants and their concentrations determine the type and concentration of the charge carriers in semiconductors. Doping lithium niobate by Fe^{3+} enhances its photorefractive property. Doping of Si by Li creates an intrinsic semiconductor and compensates for unintentional impurities. Ruby is obtained by doping sapphire by Cr^{3+}. Thallium doping in NaI helps to increase the number of light photons as output by the crystal. The literature is replete with such examples.

Specification with respect to the unintentional impurities is also equally important because these can have significant deleterious effects on the properties of the host crystal. As mentioned earlier, these may originate from the starting materials, crucibles, solvents/fluxes, and even from the furnace ambience. Hence, careful selection of the starting materials is important if one wants to prepare high-purity crystals. Wherever possible, crystals should be grown in sealed systems.

Because of their high sensitivity, most of the techniques employed for compositional analysis discussed earlier may be applied to estimate the dopant and impurity concentrations.

Compositional Inhomogeneities

Inhomogeneities in both crystal composition and dopant/impurity distribution can occur for a variety of reasons. The most important reason is the temperature fluctuations (oscillations), which lead to striations in the growing crystal. These can be detected by a number of techniques, such as light

FIGURE 12.3
Striations on (001) cleaved surface of a natural $SrSO_4$ crystal. (From Bhat, H. L., Studies on growth and defect properties of barite group crystals, PhD dissertation, SPU University, Vallabh Vidyanagar, India, 1973.)

absorption, chemical etching, impurity profiling, and so on. Figure 12.3 [6] shows an etched (001) cleaved surface of a natural $SrSO_4$ crystal where the striations are clearly seen. This crystal must have experienced abrupt fluctuations in temperature, pressure, or both during its growth.

In crystals growing on their natural habit faces, an inhomogeneous distribution of impurities may occur due to the difference in growth rates of various habit faces. This leads to preferential incorporation of impurities in certain growth sectors. Quantitative estimation of inhomogeneities will have to be carried out by techniques that allow local probing.

Morphology and Orientation

From the previous chapters, we have realized that not all growth techniques allow the crystals to grow with their natural habits. In the case of crystals with well-defined habit faces, the orientation of the crystal axes can be identified by the determination of its full morphology. In addition, if the crystal has cleavage plane(s), it will further help the identification of the crystallographic planes and directions. Therefore, determination of the morphology of grown crystals must be considered as an integral part of crystal characterization.

Goniometry

Determination of morphology is carried out by a procedure called goniometry. The instruments used for this purpose are called goniometers. They are of two kinds: contact goniometer and optical goniometer.

A simple and handy contact goniometer is shown in Figure 12.4. It consists of a semicircle head with two scales in opposing directions from 0° to 180° marked in 1-degree increments. The upper arm pivoted at the center of the graduated semicircle also sometimes features a linear scale. The method of use is to place the crystal between the straight base of the graduated semicircle (considered as an arm) and the graduated upper arm, so that the plane of the instrument is normal to the edge formed by the two faces. The recorded angle may be either the actual angle of contact or the angle between the normals of the two faces in question. Contact goniometers, although they are still useful in examining large crystals and crystals with rough surfaces, cannot be used with small crystals. Also, for accurate measurements, one uses optical goniometers.

Optical goniometers are a reflecting type. The simplest of them is a single-circle goniometer, the schematic of which is shown in Figure 12.5.

FIGURE 12.4
A simple contact goniometer.

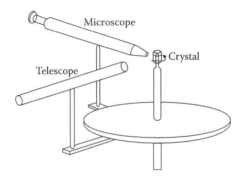

FIGURE 12.5
Schematic of a single-circle goniometer.

In this, the crystal is mounted on a rotatable vertical axis. This axis is connected to a graduated circle. The crystal is positioned such that a zone axis is parallel to the rotation axis of the graduated circle. Upon rotation of the circle, the crystal faces to be measured reflect the light from a source at specific angular positions. A telescope that is firmly attached to the base of the instrument serves as a collimator for the light source, and a microscope fixed onto a turning table receives the signal. Single-circle goniometers could be both vertical and horizontal type.

The main drawback of single-circle instruments is that they could only measure angles of the faces of one zone at a time. To measure the faces of another zone, the crystal has to be detached, remounted, and measured again. In fact, this process may have to be repeated several times in order to obtain a complete measurement of the full crystal. This problem can be avoided by using two-circle goniometers. In these instruments, the crystal is mounted on a rotatable axis, which in turn is mounted on another rotatable axis perpendicular to the first one. Thus, the positions of several crystal faces could be measured one after the other and the angular positions (φ and θ) recorded.

Crystal morphology can be generated with the help of the modern software like WinXmorph [7,8], which provides an ideal morphology with point group, lattice parameters, and upper limit of (hkl) values as inputs. Inspection of the actual morphology helps to decide which faces of the ideal morphology are to be retained. Editing the central distances (the distances of faces from the origin of the model) has to be undertaken to arrive at the actual morphology. Figure 12.6 [9] shows the actual crystal and the generated morphology of a triclinic crystal, lithium L-ascorbate dihydrate. Table 12.2 [9] gives the comparison between the experimentally determined interfacial

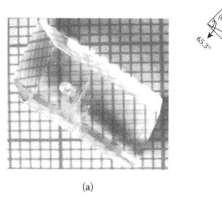
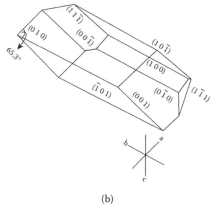

(a)   (b)

FIGURE 12.6

(a) Crystal of lithium L-ascorbate dehydrate and (b) the generated morphology. (Reprinted from *Materials Chemistry and Physics*, 137, Raghavendra Rao, K., et al., Studies on lithium L-ascorbate dihydrate: An interesting chiral nonlinear optical crystal, 756–763, 2013, with permission from Elsevier.)

TABLE 12.2

Theoretical and Experimentally Measured Interfacial
Angles in Degrees

Sl. No.	Faces	Interfacial Angle	
		Theoretical	Experimental
1	(010) & (10-1)	114.7	114.8
2	(_101) & (001)	65.3	65.2
3	(_101) & (100)	134.7	135.0
4	(_101) & (10-1)	180.0	180.0
5	(00-1) & (10-1)	65.0	65.1
6	(00-1) & (100)	110.6	110.4
7	(100) & (001)	69.4	69.1
8	(_101) & (0-10)	93.3	93.7
9	(10-1) & (0-10)	86.7	87.0
10	(001) & (0-10)	102.3	101.9
11	(-101) & (11-1)	132.0	132.3
12	(10-1) & (11-1)	48.0	48.0
13	(1-11) & (001)	56.8	56.8
14	(1-11) & (0-10)	60.8	60.7

Source: Reprinted from *Materials Chemistry and Physics*, 137,
Raghavendra Rao, K., et al., Studies on lithium L-ascorbate
dihydrate: An interesting chiral nonlinear optical crystal,
756–763, 2013, with permission from Elsevier.

angles with those theoretically estimated. As can be seen from Table 12.2, theoretical and experimental angles match very well. The morphology indicates that the most prominent faces observed in the crystal belong to [010] zone axis and that the crystal is slightly elongated along this axis.

Crystal drawings prepared in perspective as shown earlier, although quite useful, are not suitable for showing the angular relationship between habit planes and crystallographic directions. Also, in a growing crystal, sizes and shapes of habit faces are merely incidental. They often deviate quite considerably from that shown in an ideal drawing to the extent of obscuring the true symmetry relationship. In many of the calculations, we need interplanar angles rather than the size and shape of the habit planes. This requirement is met by what is known as stereographic projection. For further details on the properties and use of stereographic projections, readers are advised to refer to classical books on crystallography [10,11].

x-Ray Diffraction

If the crystal is grown however in a constrained fashion as in most high-temperature melt and vapor phase techniques, the determination of crystallographic orientation is usually done with the help of x-ray techniques. Of course, when a crystal is grown on an oriented seed, the boule axis is predetermined, but the other axes are still to be determined.

Let us take again by now the familiar example of LiNbO$_3$. For the fabrication of surface acoustic wave devices, one usually employs YZ-LiNbO$_3$ substrates wherein the acoustic waves propagate on the Y surface in the z direction. In a z axis pulled crystal, x and y axes are still to be identified. The identification of x and y axes with respect to the crystal boule can be done using the x-ray diffraction (XRD) technique. The z axis pulled LiNbO$_3$ crystals are cylindrical and usually have three growth ridges on the cylindrical surface. The ridges are parallel to the pull direction with an angular separation of 120° (in accordance with the space group R3c of the crystal). From the stereographic projection of the crystal on the z plane, one can visualize that the y axis is on one of the three ridges. To uniquely determine the y axis, XRD patterns on plates perpendicular to each of the ridge normal are recorded in the range of 20°–130°. The y plate gives only one prominent peak at 62.4° (Figure 12.7) [12]. The usefulness of both XRD and stereographic projection is illustrated in this example.

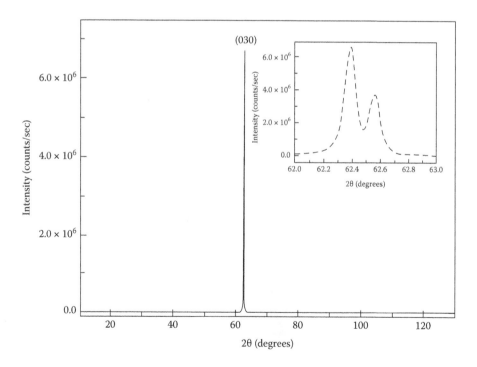

FIGURE 12.7
XRD pattern on a Y-cut LiNbO$_3$ plate. (From Vanishri, S., Studies on growth and physical properties of certain nonlinear optical and ferroelectric crystals, PhD dissertation, Indian Institute of Science, Bangalore, India, 2006.)

Dislocation Content

Measurement of dislocation density in a grown crystal gives valuable information regarding its perfection. Density of dislocation depends on many growth-related factors. The normal values may vary widely and lie anywhere between 10^0 and $10^6/cm^2$ and sometimes even higher. In the literature, examples of growth of dislocation-free crystals can also be found. In the following, we discuss a few commonly used techniques to study dislocations.

Chemical Etching

Chemical etching is the most widely used technique for studying defects in crystals, particularly the dislocations. The method consists of simply immersing a crystal having well-defined surface in a suitable medium (i.e., a liquid, a solution, or even a gaseous chemical reagent) for a certain period and then observing the surface under the microscope after proper cleaning. The medium used for this process is called by the general name "etchant." Since all structural defects are strained regions wherein some extra energy is localized, the activation energy required to remove atoms/molecules by dissolution here is reduced. Therefore, rapid nucleation of steps and their motion would occur around the emergent point of a dislocation on the crystal surface, resulting in a microscopic depression called an etch pit. Figure 12.8 shows typical etch pits produced on (001) cleavage plane of a $SrSO_4$ crystal [6].

FIGURE 12.8
Typical etch pattern on the (001) cleavage plane of $SrSO_4$ crystal (350×). (From Bhat, H. L., Studies on growth and defect properties of barite group crystals, PhD dissertation, SPU University, VallabhVidyanagar, India, 1973.)

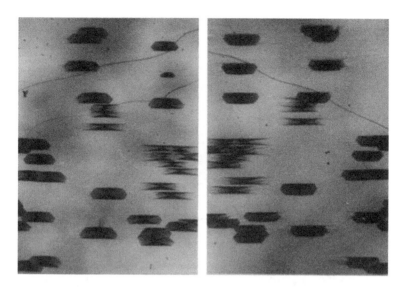

FIGURE 12.9
Etch pits on (001) cleaved match pair of $SrSO_4$ (200×). (From Bhat, H. L., Studies on growth and defect properties of barite group crystals, PhD dissertation, SPU University, Vallabh Vidyanagar, India, 1973.)

Because there is no hard and fast rule for the selection of a suitable etchant, it is usually done by the trial-and-error method. However, there are several ways to prove whether a given etchant is capable of revealing dislocation sites. For example, if etching is carried out on matched cleavage surfaces, the etch pit patterns produced on them should be a mirror image of one another, as shown in Figure 12.9 [6]. Alternately, if a crystal surface is successively etched for different durations, the number and position of the etch pits produced on the surface should more or less remain the same with only the pit size becoming progressively larger with the duration of etching. Readers may refer to the book written by Sangwal [13] to learn more about the etching technique and its capabilities.

Decoration Technique

One of the drawbacks of etch pit studies is that it is limited to surface examination. Hence, when it comes to viewing dislocations in three dimensions (i.e., dislocation configuration), it is not very useful. The decoration technique was invented to solve this problem, albeit in a limited way. In the case of crystals that are transparent to visible and infrared lights, it is possible to view dislocations in them by decorating them (hence the name). The technique consists of inducing precipitation of certain metals along the line of dislocation. The test crystal is subjected to a thermal treatment in the presence of a metal vapor (e.g., gold, silver, or copper) during which the

FIGURE 12.10
Network of dislocations in KCl crystal seen by the decoration technique. (Reprinted from *Acta Metallurgica*, 6, Amelinckx, S., Dislocation patterns in potassium chloride, 34–58, 1958, with permission from Elsevier.)

atoms of the metal diffuse into the crystal through the core of dislocations. Alternatively, the crystal may be initially doped with metal halide, as was done in the case of KCl [14], and then subjected to heat treatment. On cooling the crystal, these atoms precipitate along the dislocation lines. When viewed under the microscope in dark field, the precipitates appear like rows of beads along fine threads (Figure 12.10) [14]. The position of dislocation is revealed by the scattering of light from the precipitates. In opaque crystals like silicon, dislocations can be viewed with infrared light [15]. However, since the diffusion rates are usually small, the information one gets in the third dimension is quite limited. Also, the technique cannot be applied to opaque crystals.

x-Ray Topography

Although by no means a new technique [16], the use of x-ray topography for defect characterization has increased enormously in recent years. Since x-ray topography is an imaging technique based on the differences in reflecting power between the perfect and distorted part of the crystal, it is very suitable for studying a wide variety of defects, such as dislocations, stacking faults, domain walls, growth striations, large precipitates, growth sector boundaries, lattice strain, and so on. However, the resolution of x-ray topographic images is low and not better than about 1–5 μm. Most modern XRD topographic cameras use characteristic radiation and follow the principles laid down by Berg and Barrett for reflection and Lang for transmission modes.

Berg–Barrett Technique

Figure 12.11 [17] schematically represents the Berg–Barrett camera [18,19]. Here, the characteristic incident radiation and the active set of reflection planes are so chosen to provide a small angle of incidence a diffracted beam direction, which is determined by twice the Bragg angle is nearly normal to the sample surface. In this way, there is maximum illumination of the sample. For better resolution, the photographic plate is placed as close to the sample as physically possible. In the reflection geometry, it is always the outer skin of the sample, which is imaged typically 2–10 μm and occasionally down to a depth of 50 μm. Reflection topography is therefore particularly well suited for the studies of defects in epitaxial layers. Although the method is not suitable for bulk specimens, it is appealing because the instrumentation is simple and inexpensive.

Lang Technique

The majority of the topographic work being carried out today is with the Lang technique [20]. Lang camera design, although simple to represent schematically (Figure 12.12) [17], is considerably more sophisticated in construction than the Berg–Barrett camera. In this technique, the x-ray beam, typically consisting of a characteristic line, is passed through a narrow slit at the end of a collimating tube so that the horizontal divergence is limited to about 4–5 minutes of arc. The vertical divergence of the incident beam is controlled by a point focus source of x-ray and adjusting the distance between

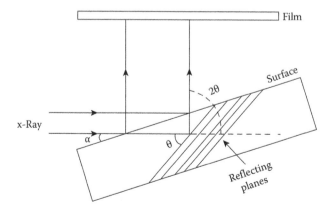

FIGURE 12.11
Schematic showing the Berg–Barrett technique. (Reprinted from *Progress in Crystal Growth and Characterization*, 11, Bhat, H. L., X-ray topographic assessment of dislocations in crystals grown from solution, 57–87, 1986.)

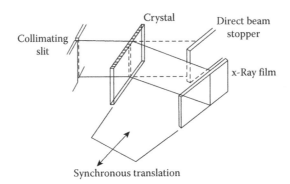

FIGURE 12.12
Geometry of transmission Lang topography. (Reprinted from *Progress in Crystal Growth and Characterization*, 11, Bhat, H. L., X-ray topographic assessment of dislocations in crystals grown from solution, 57–87, 1986.)

source and sample. The crystal mounted on a goniometer head is so oriented that the lattice planes nearly normal to the entrance and exit faces of the crystal will diffract, in which case the diffraction geometry will be symmetric. Other diffracting planes may also be chosen; however, they lead to asymmetric effects and distort the image. The diffracted beam emerging from the rear surface of the specimen passes through a slit and is recorded as a strip of blackening on the photographic plate. This is called section topograph. A complete topograph of the entire crystal is taken by activating a scanning mechanism that moves the sample across the incident beam, thereby projecting the integrated signal from the crystal onto the synchronously translating photographic plate. The crystal may have to be exposed for 2–10 h, depending on the intensity of the x-ray sources. The resulting image is called a projection or Lang x-ray topograph. A typical Lang topograph of a crystal grown from solution is shown in Figure 12.13 [21]. This figure depicts the basic features of dislocation generation and distribution in a solution-grown crystal using a seed. It contains features such as seed–crystal interface, inclusions, dislocations, growth sector boundaries, refraction of dislocation across the growth sector boundary, and so on. In x-ray topography, since the complete crystal can be mapped, it has the added advantage of giving certain information about the growth history. For example, from the zigzag shape of the growth sector boundary shown in Figure 12.13, one can infer that there must have been fluctuation in the growth rate, which must have occurred due to temperature fluctuation.

Determination of the Burgers vector of dislocations is based on image contrast (visibility rules) using $\mathbf{g.b}$ criterion, where $\mathbf{g}$ is the diffraction vector. This requires imaging of dislocations in several reflections and in the best of the cases two reflections. For example, in the case of pure screw dislocation, there will be minimum contrast for $\mathbf{g.b} = 0$ and maximum contrast for $\mathbf{g}//\mathbf{b}$. This contrast effect is shown in Figure 12.14a and b, wherein a dislocation marked D is fully visible in the former and completely invisible in

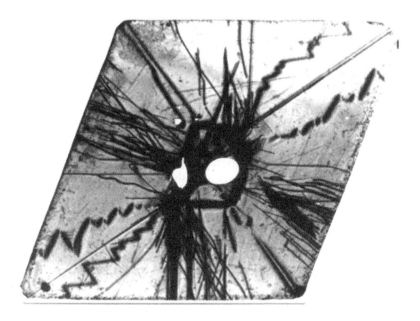

FIGURE 12.13
Distribution of defects in potash alum crystal grown over a seed. (Reprinted from *Journal of Crystal Growth*, 121, Bhat, H. L., et al., Dislocation characterization in crystals of potash alum grown by seeded solution growth under conditions of low supersaturation, 121, 709–716. 1992, with permission from Elsevier.)

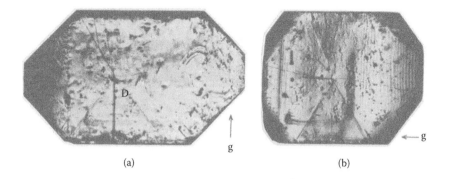

(a)   (b)

FIGURE 12.14
Image contrast for a pure screw dislocation. (a) 200 reflection and (b) 011 reflection. (Reprinted from *Progress in Crystal Growth and Characterization*, 11, Bhat, H. L., X-ray topographic assessment of dislocations in crystals grown from solution, 57–87, 1986, with permission from Elsevier; reprinted from *Journal of Applied Crystallography*, 16, Bhat, H. L., et al., The growth and perfection of crystals of ammonium dihydrogen orthophosphate, 390–398, 1981, with permission from Elsevier.)

the latter [17,22]. Similarly, it can be shown that for a pure edge dislocation the invisibility conditions are $g.b = 0$ and $g.bxl = 0$. A mixed dislocation, for which b is neither parallel nor perpendicular to l, will never be completely invisible; hence, the Burgers vector determination becomes difficult. However, there are methods of determining the Burgers vector for straight dislocations (screw, edge, and mixed) based on minimum energy and zero force theorems [23].

Burgers vector analysis in melt-grown crystal is quite involved, although the basis is still the image contrast. However, the dislocation lines here need not be straight, and one cannot easily relate them to the growth direction. One may have to take many topographs and compare the experimentally observed contrast of the images with the theoretically expected values and arrive at a result that is self-consistent.

Exposing a single crystal to an x-ray beam that has undergone a prior Bragg reflection from a highly perfect crystal(s) provides a sensitive topographic technique for measuring lattice misorientations and lattice parameter differences. A variety of geometries have been used for this technique.

Early works on x-ray topography have been reviewed by Authier [24] and Tanner [25]. For state-of-the-art information on generation and propagation of dislocations in crystals grown on habit planes, readers should refer to the chapter written by Klapper in the *Springer Handbook of Crystal Growth* [26].

Electron Microscopy

The principle of electron diffraction is broadly the same as that of XRD, where we treat the electron beam as a de Broglie wave with wavelength corresponding to the electron energy. Figure 12.15 [27] shows a typical

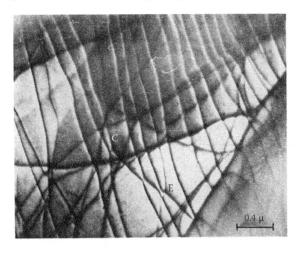

FIGURE 12.15
Electron micrograph of dislocations in cadmium iodide. (Reprinted from *Journal of Crystal Growth*, 15, Prasad, R., On the growth characteristics in cadmium iodide crystals, 259–262, 1972, with permission from Elsevier.)

electron microscopy image of dislocations wherein dislocations are imaged as dark lines. The techniques of x-ray topography and transmission electron microscopy in a way are complementary to each other in that x-ray topography enables a thick, fairly perfect crystal to be examined with relatively low resolution over a large area, whereas electron microscopy necessarily uses thin specimens of quite high dislocation density and examines only a very small area but with excellent resolution. Also, high energy of the electron often leads to melting or decomposition of nonmetallic samples, which are mostly thermally nonconducting. Consequently, the technique is extensively used to map dislocations in metals and alloys rather than nonmetals.

Low-Angle Grain Boundaries

As mentioned in Chapter 4, a low-angle grain boundary is an equally spaced array of dislocations, and its presence in a crystal in large numbers results in mosaic structure. Their presence consequently affects the mechanical properties of the crystal like yield, creep, plastic deformation, and fracture. Crystal growers normally would like to bring their concentration to the minimum. Because of their association with dislocations, their detection in the crystal is facilitated by the techniques that reveal dislocations. For example, chemical etching of the crystal results in equally spaced etch pits along the low-angle grain boundary (Figure 12.16a) [6].

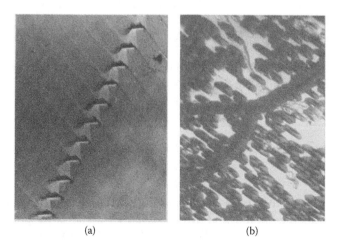

(a)   (b)

FIGURE 12.16
Low-angle grain boundaries: (a) grain boundary in high magnification showing regular spacing between etch pits (2000×) and (b) intersecting grain boundaries (200×). (From Bhat, H. L., Studies on growth and defect properties of barite group crystals, PhD dissertation, SPU University, VallabhVidyanagar, India, 1973.)

Boundaries are rarely observed in isolation but all over the crystal. Tilt boundaries often intersect in the form of T's or L's; a T intersection is shown in Figure 12.16b.

Twins

Twinned crystal is a composite crystal in which its components, separated by a boundary, are related to each other in a simple crystallographic manner. Twin boundaries occur when two parts of the crystal intergrow so that only a slight misorientation exists between them. It is a highly symmetrical interface, and often the adjacent parts of the crystal would be the mirror image of one another. The boundary is also a much lower energy interface than the grain boundaries that form when crystals of arbitrary orientation grow together. Consequently, they are, as a rule, densely packed low-index planes common to the twin partners.

The twins can occur during growth as well as after growth, the latter being due to deformation. These are respectively called growth twins and deformation twins. Crystal growers are naturally worried about the former and would like to avoid their formation. Growth twins are formed in the early stages of the growth itself, as evidenced by the fact that the twin parts are of similar size and appear to have originated from one common point, which is usually the center of the twinned crystal from where it started growing.

In the case of transparent crystals, their identification can be done by observing them with crossed polarizers. However, for opaque crystals, detailed x-ray examination would be required, although simple chemical etching can often delineate the twinned regions.

Inclusions

Inclusions are another type of defects that should be avoided. These are regions within the crystal that are occupied by some other phase, which may be solid, liquid, or gas. For example, crystals growing from solution often trap mother liquor in the form of shallow veils parallel to the growth face. Usually the cavities enclosing the inclusions are of varying sizes and shapes and may be distributed randomly in a crystal. But often they are arranged in a preferential fashion. If the growing surfaces are

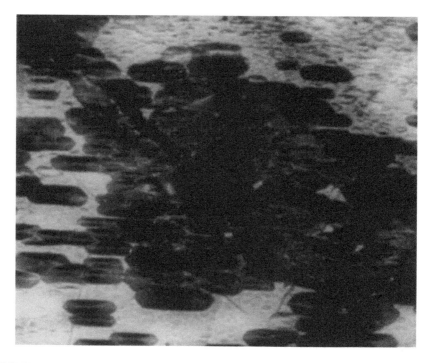

FIGURE 12.17
Etch pits around an inclusion. (From Bhat, H. L., Studies on growth and defect properties of barite group crystals, PhD dissertation, SPU University, VallabhVidyanagar, India, 1973.)

rounded due to dissolution, facets of the habit faces are formed over them during the initial stage of growth leading to capping. The capping process favors entrapment of solvent inclusions.

Formation of inclusion depends on a number of growth parameters, such as temperature fluctuations, growth rate, composition of the parent solution, and so on. The hydrodynamics of the solution flow around a growing crystal would also play an important role in the formation of solvent inclusions. These inclusions often act as a source of dislocations, which in turn affect the mechanical properties of the crystal. Figure 12.17 [6] shows an inclusion around which a large number of dislocations are seen, as revealed by chemical etching.

If all the basic parameters discussed earlier are known for a given crystal, the end user will have a reasonable knowledge of the crystal and can proceed further in its application. However, depending on the end use, additional tests may be required that will have to be performed at the user end. Nevertheless, it is important to realize that greater emphasis should be given for postgrowth evaluation of the crystals than what is normally perceived.

References

1. Elwell, D., and H. J. Scheel. 1975. *Crystal growth from high temperature solution.* London, UK: Academic Press.
2. Brice, J. C. 1968. *Crystal growth processes.* London, UK: Blackie and Sons.
3. Madaro, F., R. Sæterli, J. R. Tolchard, M.-A. Einarsrud, R. Holmestad, and T. Grande. 2011. Molten salt synthesis of $K_4Nb_6O_{17}$, $K_2Nb_4O_{11}$ and KNb_3O_8 crystals with needle- or plate-like morphology. *Crystal Engineering Communications* 13: 1304–1313.
4. Brahadeeswaran, S., V. Venkataramanan, J. N. Sherwood, and H. L. Bhat. 1998. Crystal growth and characterization of semi-organic nonlinear optical material: Sodium p-nitrophenolate dehydrate. *Journal of Material Chemistry* 8: 613–618.
5. Dixit, V. K. 2003. Bulk and thin film growth of pure and substituted indium antimonide for infrared detector applications. PhD dissertation, Indian Institute of Science, Bangalore, India.
6. Bhat, H. L. 1973. Studies on growth and defect properties of barite group crystals. PhD dissertation, SPU University, VallabhVidyanagar, India.
7. Kaminsky, W. 2005. WinXMorph: A computer program to draw crystal morphology, growth sectors and cross sections with export files in VRML V2.0 utf8-virtual reality format. *Journal of Applied Crystallography* 38: 566–567.
8. Kaminsky, W. 2007. From CIF to virtual morphology using WinXMorph program. *Journal of Applied Crystallography* 40: 382–385.
9. Raghavendra Rao, K., H. L. Bhat, and E. Suja. 2013. Studies on lithium L-ascorbate dihydrate: An interesting chiral nonlinear optical crystal. *Materials Chemistry and Physics* 137: 756–763.
10. Phillips, F. C. 1971. *An introduction to crystallography,* 4th ed. New York, NY: ELBS and Longman Group.
11. Azaroff, L. V. 1968. *Elements of crystallography.* New York, NY: McGraw-Hill.
12. Vanishri, S. 2006. Studies on growth and physical properties of certain nonlinear optical and ferroelectric crystals. PhD dissertation, Indian Institute of Science, Bangalore, India.
13. Sangwal, K. 1987. *Etching of crystals: Theory, experiment and application: Defects in solids,* Vol. 15. Amsterdam, The Netherlands: North Holland.
14. Amelinckx, S. 1958. Dislocation patterns in potassium chloride. *Acta Metallurgica* 6: 34–58.
15. Dash, W. C. 1956. Copper precipitation on dislocations in silicon. *Journal of Applied Physics* 27: 1193–1195.
16. Ramachandran, G. N. 1944. X-ray topographs of diamond. *Proceedings of the Indian Academy of Sciences A* 19: 280–292.
17. Bhat, H. L. 1986. X-ray topographic assessment of dislocations in crystals grown from solution. *Progress in Crystal Growth and Characterization* 11: 57–87.
18. Berg, W. L. 1931. Über ein röntgenographische Methode zur von Gitterstörungen an Kristallen. *Naturwissenschaften* 19: 391–396.
19. Barrett, C. S. 1931. Laue spots from perfect, imperfect and oscillating crystals. *Physics Review* 38: 832–833.
20. Lang, A. R. 1959. A projection topograph: A new method in X-ray diffraction micro radiography. *Acta Crystallographica* 12: 249–250.

21. Bhat, H. L., R. I. Ristic, J. N. Sherwood, and T. Shripathi. 1992. Dislocation characterization in crystals of potash alum grown by seeded solution growth under conditions of low supersaturation. *Journal of Crystal Growth* 121: 709–716.
22. Bhat, H. L., K. J. Roberts, and J. N. Sherwood. 1981. The growth and perfection of crystals of ammonium dihydrogen orthophosphate. *Journal of Applied Crystallography* 16: 390–398.
23. Klapper, H. 1996. Defects in non-metal crystals. In *Characterization of crystal growth defects by X-ray methods*, edited by B. K. Tanner and D. K. Bowen. New York, NY: Plenum Press, pp. 133–160.
24. Authier, A. 1972. X-ray topography as a tool in crystal growth studies. *Journal of Crystal Growth* 13/14: 34–38.
25. Tanner, B. K. 1976. *X-ray diffraction topography*. Oxford, UK: Pergamon Press.
26. Klapper, H. 2010. Generation and propagation of defects during crystal growth. In *Springer handbook of crystal growth*, edited by G. Dhanaraj et al. New York, NY: Springer, pp. 93–132.
27. Prasad, R. 1972. On the growth characteristics in cadmium iodide crystals. *Journal of Crystal Growth* 15: 259–262.

Appendix I

Mobility of Face-Adsorbed Molecules

Here, we estimate the mean diffusion distance of a face-adsorbed molecule. Assuming that the face-adsorbed molecules (labeled A in Figure 3.2a) are essentially mobile, the growth process of a crystal surface with steps can be considered to be the result of three sequential steps: (1) exchange of molecules between the adsorbed layer and the vapor, (2) diffusion of adsorbed molecules toward the step, and (3) possible diffusion of adsorbed molecules at the edge of the steps toward the kinks and attachment at this site.

To discuss the role of diffusion on the surface, let us introduce the mean displacement x_s of face-adsorbed molecules. This can be defined in general terms using Einstein's diffusion formula,

$$x_s^2 = D_s \tau_s \tag{AI.1}$$

where D_s is the diffusion coefficient and τ_s is the mean lifetime of an adsorbed molecule before getting evaporated back into the vapor. For simple molecules, we can write

$$D_s = a^2 \nu' e^{-U_s/kT} \tag{AI.2}$$

and

$$1/\tau_s = \nu' e^{-W_s'/kT} \tag{AI.3}$$

Here, U_s is the activation energy between two adjacent equilibrium sites on the surface, at a distance a from each other, and W_s' is the energy of evaporation from the surface to the vapor. The frequency factors ν' and ν would both be of the order of atomic frequency of vibration ($\nu \sim 10^{13}$ s^{-1}) in the case of monoatomic crystals, but they will be different in the case of more complicated molecules. Hence, assuming $\nu \sim \nu'$, and substituting Equations AI.2 and AI.3 in Equation AI.1, we have

$$x_s = a e^{(W_s' - U_s)/2kT} \tag{AI.4}$$

If the diffusion on the surface has to play an important role, $x_s > a$, and therefore $W_s' > U_s$. This is probably always the case and therefore x_s can be much larger than a. Also, it increases rapidly as the temperature decreases.

To get an idea of the values that we can expect for x_s, let us consider a (111) close-packed surface of a face-centered cubic (fcc) crystal. By simple consideration and assuming only the nearest neighbor interaction, we have $W_s' = 3\phi$, which is half the total evaporation energy $W = W_s + W_s'$ while $U_s \sim \phi = \dfrac{1}{6}W$.

However, detailed calculations carried out by Mackenzic [1] using Lennard-Jones forces show that U_s is considerably smaller and is about $1/20\ W$. In this case, x_s becomes

$$x_s \sim a e^{\,3\phi/2kT} \sim 4 \times 10^2\, a \tag{AI.5}$$

for a typical value $\phi/kT \sim 4$.

It may be noted that x_s will be a function of the crystal face considered because both W_s' and U_s will be different for different faces. In general, x_s will be the smallest for the closest packed surface because W_s' increases more rapidly than U_s. Also, a is very small for such a surface. For instance, for a (100) face in fcc crystal, assuming nearest neighbor forces only, $W_s' \approx 4\phi$ and U_s is probably still very small. Thus,

$$x_s \sim a e^{\,2\phi/kT} \sim 3 \times 10^3\, a \tag{AI.6}$$

Reference

1. Mackenzie, J. K. 1949. A theory of sintering and the theoretical yield strength of solids. A PhD dissertation, University of Bristol, Bristol.

Appendix II

Concentration of Kinks in a Step

Frenkel [1] was the first to point out that a monomolecular step on a crystal surface contains a high concentration of kinks. Burton and Cabrera [2] later showed that the concentration of kinks is even larger than that proposed by Frenkel. This result is very important from the point of view of the rate of advance of the steps. We present now Burton and Cabrera's approach to determine the concentration of kinks in a step.

Let a close-packed crystallographic direction be taken as the x axis and consider a step that follows this axis in the mean, so that the surface is one molecule higher in the region $y < 0$ than in the region $y > 0$ (Figure AII.1). Following the step along the positive x direction, the points where y increases or decreases by a unit spacing a are called positive or negative kinks, respectively. We assume that this step contains an equal number of positive and negative kinks. Let $2n$ and q be the probabilities for having a kink or no kink, respectively, at a given site in the step. Then, we can write

$$\frac{n}{q} = e^{-w/kT} \quad \text{and} \quad 2n+q = 1 \tag{AII.1}$$

where w is the energy necessary to form a kink. Now, the mean distance between the kinks is $x_0 = a/2n$, where a is the intermolecular distance in the direction of the step. We can derive an expression for x_0 in the following way.

From Equation AII.1, we have

$$q = ne^{w/kT} \quad \text{or} \quad 2n + ne^{w/kT} = 1 \quad \text{or} \quad e^{w/kT} + 2 = 1/n \tag{AII.2}$$

Further

$$x_0 = \frac{a}{2n} = \frac{a}{2}(e^{w/kT} + 2) \sim \frac{a}{2}e^{w/kT}$$

Therefore

$$x_0 \sim \frac{1}{2}a \, \exp(w/kT) \tag{AII.3}$$

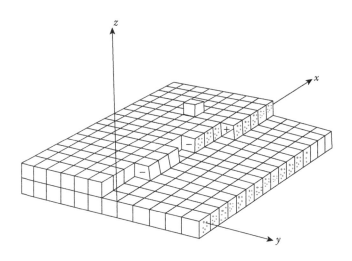

FIGURE AII.1
Positive (marked +) and negative (marked –) kinks.

Again, as the inclination of the step relative to a close-packed direction increases, the number of kinks increases and the mean distance $x_0(\theta)$ in that case can be written as

$$x_0(\theta) = x_0 \left\{ 1 - \frac{1}{2} \left(\frac{x_0}{a} \right)^2 \theta \right\} \qquad \text{(AII.4)}$$

Detailed calculations of w on several crystal surfaces are available in the literature [3–5]. However, simple considerations suggest that w must be a small fraction of evaporation energy. For instance, in a close-packed step on a (111) face of a face-centered cubic crystal with nearest neighbor interactions, it is easy to see that w is equal to a quarter of the energy necessary to move one molecule from a position in the step to an adsorption position against the straight step (equal to 2ϕ); hence, $w = (\tfrac{1}{2})\phi, = 1/12\ W$ (because $\phi = 1/6\ W$).
The mean distance between the kinks therefore becomes

$$x_0 = \frac{1}{2} a e^{(\phi/2kT)} \qquad \text{(AII.5)}$$

For growth under typical conditions, which may be taken such that $(\phi/kT) \sim 4$,

$$x_0 \approx 4a \qquad \text{(AII.6)}$$

Under these conditions we shall have a kink for every four sites in the step.

References

1. Frenkel, J. 1945. Viscous flow of crystalline bodies under the action of surface tension. *Journal of Physics USSR* 9: 385–391.
2. Burton, W. K., and N. Cabrera. 1949. Crystal growth and surface structure. Part I. *Discussions of the Faraday Society* 5: 33–39.
3. Bikov, A. Z., S. D. Stoichlov, and S. B. Damianova. 1994. Distribution of the gibbs activation energy for formation of a kink around the top of the growth hills of the face (001) of TGS single crystal. *Crystal Res and Technology* 29: 13–17.
4. Vitos, L., H. L. Skriver, and J. Kollar. 1999. The formation energy for steps and kinks on cubic transition metal surfaces. *Surface Science* 425: 212–223.
5. Feibelman, P. J. 2000. Step- versus kink-formation energies on Pt(111). *Surface Science* 463: L661–L665.

Appendix III

Rate of Advancement of a Straight Step

Due to accretion of the molecules, a straight step will advance. As mentioned in Chapter 3 the points on the straight step where growth actually occurs are the kinks in the step. This can be pictured again in the two dimensional layer growth model, for when an atom meets a step, it is held at the step where it is more tightly bound to the crystal being in contact on two of its sides. However, it continues to diffuse along the step until it hits a kink where it makes three bonds with the crystal. From these considerations, one can derive a formula for rate of advance of the step and see what influences this rate. We will actually see that this rate is proportional to the supersaturation.

As mentioned in Chapter 2, if the actual vapor pressure is p and the saturation value is p_0, the saturation ratio α and the supersaturation S are defined as

$$\alpha = p/p_0 \quad \text{and} \quad S = \alpha - 1 = (p/p_0) - 1 = \frac{p - p^\circ}{p^\circ} = \Delta \, p/p^\circ \qquad \text{(AIII.1)}$$

The surface concentration of face-adsorbed molecules at a large distance from the growing step is $n_\alpha = n_0 \alpha$, whereas we may assume the concentration at the step itself to be the equilibrium value given by

$$n_0 = \frac{1}{a^2} e^{-W_s/kT} \qquad \text{(AIII.2)}$$

This is valid if we assume that kinks on the growing edge to be in equilibrium with a surface concentration of the adsorbed molecules n_0, which in simple cases will be of the order of number of molecular positions per centimeter square. We remind ourselves that W_s is the energy of evaporation of a molecule from the kink site to a site on the surface.

A reasonable estimate of the smallest distance from the step measured along the surface at which the concentration is n_α is the mean diffusion distance x_s of the adsorbed molecule, giving rise to a concentration gradient of
$$\frac{n_0 \alpha - n_0}{x_s} = \frac{n_0(\alpha - 1)}{x_s}.$$

Now, v_∞, the rate of advance of a straight step, is a^2 times the number of molecules arriving at 1 cm of the edge per second. Thus, we have

$$v_\infty = a^2 . 2 . D \frac{n_0(\alpha - 1)}{x_s} \tag{AIII.3}$$

The factor 2 arises because the molecules diffuse from both the directions toward the step.

Earlier we had seen that $x_s^2 = D \, \tau_s$ and $1/\tau_s = v e^{(-W_s'/kT)}$

Therefore,

$$D = x_s^2 \, v e^{(-W_s/kT)} \tag{AIII.4}$$

Substituting the value of D in Equation AIII.3, we get

$$v_\infty = a^2 \, 2 . x_s^2 \, v e^{(-W_s'/kT)} \frac{n_0(\alpha - 1)}{x_s} \tag{AIII.5}$$

$$= 2 \, (\alpha - 1) \, x_s \, v e^{(-W_s'/kT)} \, n_0 a^2 \tag{AIII.6}$$

$$= 2 \, (\alpha - 1) \, x_s \, v e^{-W/kT} \tag{AIII.7}$$

because $n_0 \, a^2 = e^{-W_s/kT}$ and $W = W_s + W_s$.

Qualitatively, this would imply that all molecules that hit the surface in the diffusion zone of width $2x_s$ will reach the advancing step, and because there is a large concentration of kinks in the step, they will be adsorbed.

A curved step with radius of curvature ρ advances with a velocity v_ρ given by

$$v_\rho = v_\infty \, [1 - (\rho_c/\rho)] \tag{AIII.8}$$

where ρ_c is the critical radius of curvature.

The equations for rate show that the velocity with which the edge of the layer travels is proportional to supersaturation $S = \alpha - 1 = \Delta p/p_0$.

Appendix IV

Refinement in Calculation of Critical Shear Stress

The critical shear strength depends upon the hardness of the atoms—that is, upon the steepness with which the repulsive forces between the atoms builds up as they are brought together. To see this, consider the following case, where an atom A is taking part in the process of slip by sliding over its neighbor B in the lower row (Figure AIV.1a).

This would happen when the applied force per atom is large enough to overcome the bond between atoms A and C. A detailed analysis of this problem would be involved, but the principle of this argument can be illustrated by means of a simpler problem, where two atoms in a molecule are being separated as in Figure AIV.1b. Let the force between these atoms be represented by

$$F = K\left[\left(\frac{r_0}{r}\right)^n - \left(\frac{r_0}{r}\right)^m\right] \tag{AIV.1}$$

where K is a constant, r is the spacing between the atoms, r_0 is the equilibrium spacing, and the index m of the repulsive component of forces is greater than n, that of the attractive component. Equating dF/dr to zero, we can find the spacing r at which F is a maximum.

$$\frac{dF}{dr} = 0 = K\left[-n\frac{r_0^n}{r^{n+1}} - -m\frac{r_0^m}{r^{m+1}}\right] \quad \text{or} \quad \left[n\frac{r_0^n}{r^{n+1}} = m\frac{r_0^m}{r^{m+1}}\right] \tag{AIV.2}$$

$$\frac{m}{n} = \frac{r_0^n}{r^{n+1}} \cdot \frac{r^{m+1}}{r_0^m} = \frac{r^{m-n}}{r_0^{m-n}} \tag{AIV.3}$$

Therefore,

$$\frac{r}{r_0} = \left(\frac{m}{n}\right)^{1/m-n} \quad \text{or} \quad r = r_0\left(\frac{m}{n}\right)^{1/m-n} \tag{AIV.4}$$

The particular values of m and n to be used depend upon the nature of the atoms involved. For soft compressible atoms, $m = 4$ would be reasonable. Further, if we assume $n = 2$ (inverse square attraction), then $r/r_0 \approx 1.4$.

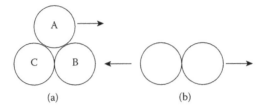

FIGURE AIV.1
Slip of an atom A over the lower plane being approximated to separation of two atoms.

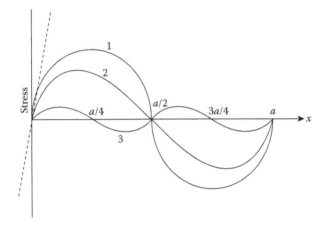

FIGURE AIV.2
Dependence of shear stress as a function of relative displacement after refinement.

When repulsion is mainly due to overlapping of closed shells, as in full metals (Cu) and ionic salts, the atoms are less compressible and larger values of m have to be used, rising in extreme cases to $m = 12$. Taking $m = 12$ and $n = 2$, we get $r/r_0 \approx 1.2$ for $m = 12$ and $n = 7$ (van der Waal's attraction) $r/r_0 \approx 1.11$. Thus, the critical shear strain that must be exceeded to break the bond is smaller for hard atoms than soft ones. On the other hand, it is not possible to reduce the critical strain to less than a few percent without assuming unreasonable values of m and n. Thus, the sinusoidal relation shown by curve 1 in Figure AIV.2 may be replaced by a more realistic curve of type 2.

Another point to be noted is that the calculation of theoretical stress takes no account of possible configuration of mechanical stability through which the lattice may pass as it is sheared. The face-centered cubic metals, for example, pass through twinning and then body-centered cubic configurations when sheared on their slip planes. In such cases, the force-displacement relation will oscillate over smaller periods than the lattice spacing, as in curve 3, and the critical shear stress will be further reduced.

Index